Security in IoT
Social Networks

T0318429

Security in IoT Social Networks

Edited by

Fadi Al-Turjman

*Full Professor and Research Center Director
at Near East University, Turkey*

B.D. Deebak

*Associate Professor in the Department of
Computational Intelligence, School of Computer Science
and Engineering at Vellore Institute of Technology,
Vellore, India*

Series Editor: Fatos Xhafa

Universitat Politècnica de Catalunya, Spain

Academic Press is an imprint of Elsevier
125 London Wall, London EC2Y 5AS, United Kingdom
525 B Street, Suite 1650, San Diego, CA 92101, United States
50 Hampshire Street, 5th Floor, Cambridge, MA 02139, United States
The Boulevard, Langford Lane, Kidlington, Oxford OX5 1GB, United Kingdom

Copyright © 2021 Elsevier Inc. All rights reserved.

No part of this publication may be reproduced or transmitted in any form or by any
means, electronic or mechanical, including photocopying, recording, or any information
storage and retrieval system, without permission in writing from the publisher. Details on
how to seek permission, further information about the Publisher's permissions policies
and our arrangements with organizations such as the Copyright Clearance Center and the
Copyright Licensing Agency, can be found at our website: www.elsevier.com/permissions.

This book and the individual contributions contained in it are protected under copyright
by the Publisher (other than as may be noted herein).

Notices
Knowledge and best practice in this field are constantly changing. As new research and
experience broaden our understanding, changes in research methods, professional
practices, or medical treatment may become necessary.

Practitioners and researchers must always rely on their own experience and knowledge in
evaluating and using any information, methods, compounds, or experiments described
herein. In using such information or methods they should be mindful of their own safety
and the safety of others, including parties for whom they have a professional
responsibility.

To the fullest extent of the law, neither the Publisher nor the authors, contributors, or
editors, assume any liability for any injury and/or damage to persons or property as a
matter of products liability, negligence or otherwise, or from any use or operation of any
methods, products, instructions, or ideas contained in the material herein.

Library of Congress Cataloging-in-Publication Data
A catalog record for this book is available from the Library of Congress

British Library Cataloguing-in-Publication Data
A catalogue record for this book is available from the British Library

ISBN: 978-0-12-821599-9

For information on all Academic Press publications visit our
website at https://www.elsevier.com/books-and-journals

Publisher: Mara Conner
Acquisitions Editor: Sonnini R. Yura
Editorial Project Manager: Emily Thomson
Production Project Manager: Maria Bernard
Cover Designer: Victoria Pearson

Typeset by TNQ Technologies

Working together
to grow libraries in
developing countries

www.elsevier.com • www.bookaid.org

Contents

CHAPTER 4   **Influence of social information networks**
and their propagation..**83**

B. Raja Koti, G.V.S. Raj Kumar, K. Naveen Kumar and
Y. Srinivas

**CHAPTER 7   Recommender systems: security threats and
mechanisms** **149**

*Satya Keerthi Gorripati, Anupama Angadi and
Pedada Saraswathi*

Contributors

Fadi Al-Turjman
Research Center for AI and IoT, Near East University, Nicosia, Mersin, Turkey

Anupama Angadi
Computer Science and Engineering, Raghu Engineering College (Autonomous), Dakamarri, Viskhapatnam, Andhra Pradesh, India

B.D. Deebak
School of Computer Science and Engineering, Vellore Institute of Technology, Vellore, Tamil Nadu, India

G.V. Eswar
Department of Computer Science and Engineering, Anil Neerukonda Institute of Technology and Science, Visakhapatnam, Andhra Pradesh, India

N. Gayathri
School of Computing Science and Engineering, Galgotias University, Greater Noida, Uttar Pradesh, India

Satya Keerthi Gorripati
Computer Science and Engineering, Gayatri Vidya Parishad College of Engineering (Autonomous), Visakhapatnam, Andhra Pradesh, India

S. Kathiravan
School of Computer Science and Engineering, Vellore Institute of Technology, Vellore, Tamil Nadu, India

Dasari Siva Krishna
Computer Science and Engineering, GMR Institute of Technology, Srikakulam, Andhra Pradesh, India

K. Naveen Kumar
Department of Computer Science and Engineering, GITAM Institute of Technology, GITAM (Deemed to be University), Visakhapatnam, Andhra Pradesh, India

B. Raja Koti
Department of Computer Science and Engineering, GITAM Institute of Technology, GITAM (Deemed to be University), Visakhapatnam, Andhra Pradesh, India

G.V.S. Raj Kumar
Department of Computer Science and Engineering, GITAM Institute of Technology, GITAM (Deemed to be University), Visakhapatnam, Andhra Pradesh, India

S. Rakesh kumar
School of Computing Science and Engineering, Galgotias University, Greater Noida, Uttar Pradesh, India

Patruni Muralidhara Rao
School of Computer Science and Engineering, Vellore Institute of Technology, Vellore, Tamil Nadu, India

Thammada Srinivasa Rao
Computer Science and Engineering, GITAM Institute of Technology, GITAM, Visakhapatnam, Andhra Pradesh, India

Sivaranjani Reddi
Department of Computer Science and Engineering, Anil Neerukonda Institute of Technology and Science, Visakhapatnam, Andhra Pradesh, India

Sanjiban Sekhar Roy
School of Computer Science and Engineering, Vellore Institute of Technology, Vellore, Tamil Nadu, India

Ramiz Salama
Computer Engineering Department, Research Center for AI and IoT, Near East University, Nicosia, Mersin, Turkey

Pedada Saraswathi
Department of Computer Science and Engineering, GITAM Institute of Technology, Visakhapatnam, Andhra Pradesh, India

Y. Srinivas
Department of Computer Science and Engineering, GITAM Institute of Technology, GITAM (Deemed to be University), Visakhapatnam, Andhra Pradesh, India

Deepanshu Srivastava
School of Computing Science and Engineering, Galgotias University, Greater Noida, Uttar Pradesh, India

Hadi Zahmatkesh
Department of Mechanical, Electronic and Chemical Engineering, OsloMet - Oslo Metropolitan University, Oslo, Norway

Security in social networks

1

Fadi Al-Turjman[1], Ramiz Salama[2]

[1]*Research Center for AI and IoT, Near East University, Nicosia, Mersin, Turkey;* [2]*Computer Engineering Department, Research Center for AI and IoT, Near East University, Nicosia, Mersin, Turkey*

1. Introduction

Sites are defined as interactive web-based applications (apps) that provide users with the ability to communicate with friends and family, meet new people, join groups, chat, share photos, and organize events and network with others in a similar-to-real-life manner. Security network (SN) functionalities are organized into three main categories: social networks services (SNS), network application services (NAS), and the communication interface (CI). SNS are used to establish social network relationships between people who have the same activities and interests. NAS provide network interaction services for users such as psychological tests, social web games, or fan groups. CI offers platforms to support users' communication and interaction. The privacy paradox is an interesting phenomenon that takes place in SN websites, where people are usually more protective of their personal information when using different communication media (i.e., personal or phone) compared to their readiness to provide this information via the SN websites. The Internet connects the whole world over this digital network, which makes it more difficult to protect information using traditional technical solutions. Knowing the purpose behind information theft and attacks on SN sites helps in providing the best techniques to protect the users' information [1].

Attackers and fraudsters might attack just for fun, to show that they can penetrate secure systems, and others might attack to gain control over systems to organize devices into a botnet to apply denial of service (DoS) attacks. However, the most common reason is the financial benefit gained by collecting a user's critical personal information such as bank accounts, social security numbers, and passwords. By doing so, attackers can commit identity theft crimes and generate profit. There are different precautions that should be considered beside the technical solutions. These include raising a user's awareness to help them distinguish between sensitive and public information. In addition, SN sites should play a major role in protecting personal information. They should enhance spam and malicious links filtering, notify users when any attack takes place, and program the sites carefully to be protected against platform attacks and other attacks likes the SQL injection and cross-site scripting, which can be added to the web page code to steal cookies, or force users to download malware and hijack users' accounts [2].

When it comes to privacy and security issues on social networks, "the sites most likely to suffer from issues are the most popular ones," Graham Cluley, Chief Technology Officer at UK tech security firm Sophos, says. But security issues and privacy issues are entirely two different beasts. A security issue occurs when a hacker gains unauthorized access to a site's protected coding or written language. Privacy issues, those involving the unwarranted access of private information, do not necessarily have to involve security breaches. Someone can gain access to confidential information by simply watching you type your password. But both types of breaches are often intertwined on social networks, especially since anyone who breaches a site's security network opens the door to easy access to private information belonging to any user. But the potential harm to an individual user really boils down to how much a user engages in a social networking site, as well as the amount of information they are willing to share. In other words, a Facebook user with 900 friends and 60 group memberships is a lot more likely to be harmed by a breach than someone who barely uses the site.

Security lapses on social networks do not necessarily involve the exploitation of a user's private information. Take, for example, the infamous "Samy" MySpace XSS worm that effectively shut the site down for a few days in October 2005. The Samy virus (named after the virus' creator) was fairly harmless, and the malware added the words "Samy Is My Hero" to the top of every affected user's MySpace profile page. This was a colossal inconvenience, naturally, but nobody's identity was stolen, and no private information was leaked. In the end, the problem galvanized the MySpace team to roll up their sleeves and seriously tighten the site's security. The result has been no major break-ins since. Unfortunately, these kinds of breaches, purely for sport in Samy's case, are rare.

The reason social network security and privacy lapses exist results simply from the astronomic amounts of information the sites process each and every day that ends up making it that much easier to exploit a single flaw in the system. Features that invite user participation, messages, invitations, photos, open platform applications, etc., are often the avenues used to gain access to private information, especially in the case of Facebook. Adrienne Felt, a Ph.D. Candidate at Berkeley, made small headlines last year when she exposed a potentially devastating hole in the framework of Facebook's third-party application programming interface (API) that allows for easy theft of private information. Felt and her coresearchers found that third-party platform applications for Facebook gave developers access to far more information (addresses, pictures, interests, etc.) than needed to run the app.

This potential privacy breach is actually built into the systematic framework of Facebook, and unfortunately the flaw renders the system almost indefensible. "The question for social networks is resolving the difference between mistakes in implementation and what the design of the application platform is intended to allow," David Evans, Assistant Professor of Computer Science at the University of Virginia, says. There's also the question of whom we should hold responsible for the oversharing of user data. That resolution is not likely to come anytime soon, says Evans, because a new, more regulated API would require Facebook "to break a lot of applications, and a lot of companies are trying to make money off applications now." Felt agrees, noting that now "there are marketing businesses built on top of the idea that third parties can get access to data on Facebook."

The problems plaguing social network security and privacy issues, for now, can only be resolved if users take a more careful approach to what they share and how much. With the growth of social networks, it is becoming harder to effectively monitor and protect site users and their activity because the tasks of security programmers become increasing spread out. Imagine if a prison whose inmate count jumped from a few dozen to 250 million in less than 5 years only employed 300 guards (in the case of MySpace). In response to the potential threats to which users are exposed, most of the major networks now enable users to set privacy controls for who has the ability to view their information. But, considering the application loophole in Facebook, increased privacy settings do not always guarantee privacy. Even when the flawed API was publicly exposed, "Facebook changed the wording of the user agreement a little bit, but nothing technically to solve the problem," says Evans. That means if a nefarious application developer wanted to sell the personal info of people who used his app to advertising companies, he or she could.

Yet, users still post tons of personal data on social networks without batting an eye. It is only natural. Anonymity and the fact that you are communicating with a machine instead of an actual person (or people in the case of social networking) makes sharing a lot easier. "People should just exercise common sense online, but the problem with common sense is that it's not very common. If you would not invite these people into your house to see your cat, you certainly would not let them see pictures from holiday," says Cluley.

In the end, the only tried and true solution to social network privacy and security issues is to limit your presence altogether. Do not post anything you would not mind telling a complete stranger, because in reality that is the potential for access. Be careful who you add as a "friend," because there is simply no way of verifying a user's actual identity online. Cluley compares it to a representative from your company's IT department calling to ask for your login password: "Most people will give it over" with no proof of the IT representative actually existing. The caller might be your IT representative, or she might not be. "This kind of scam happens all the time," says Cluley. Friends on social networks should know that real friends should know personal information already, negating the need to post it online.

Will there ever be a security breach-free social network? Probably not. "Any complex system has vulnerabilities in it. It is just the nature of building something above a certain level of complexity," says Professor Evans. According to Felt, the best idea is a completely private social network. "It simply requires that there's no gossip in the circle, by which I mean one person who sets their privacy settings so low that third parties can use them to get to their friends." "Social networks are great fun, and can be advantageous but people really need to understand that it's a complicated world and you need to step wisely," Cluley says.

This paper overviews significant issues related to the security of SN sites and includes the main challenges facing privacy measures. Moreover, it offers a discussion about the varying approaches in addressing these security issues.

2. **Types of social networks**

There are many types of social networks available. Most social networks combine elements of more than one of these types of networks, and the focus of a social network may change over time. Many of the security and privacy recommendations are applicable to other types of networks.

2.1 **Personal networks**

These networks allow users to create detailed online profiles and connect with other users, with an emphasis on social relationships such as friendship. For example, Facebook, Friendster, and MySpace are platforms for communicating with contacts. These networks often involve users sharing information with other approved users, such as one's gender, age, interests, educational background, and employment, as well as files and links to music, photos, and videos. These platforms may also share selected information with individuals and applications that are not authorized contacts.

2.2 **Status update networks**

These types of social networks are designed to allow users to post short status updates to communicate with other users quickly. For example, Twitter focuses its services on providing instantaneous, short updates. These networks are designed to broadcast information quickly and publicly, though there may be privacy settings to restrict access to status updates.

2.3 **Location networks**

With the advent of GPS-enabled cellular phones, location networks are growing in popularity. These networks are designed to broadcast one's real-time location, either as public information or as an update viewable to authorized contacts. Many of these networks are built to interact with other social networks, so an update made to a location network could (with proper authorization) post to one's other social networks. Some examples of location networks include Brightkite, Foursquare, Loopt, and Google Latitude.

2.4 **Content-sharing networks**

These networks are designed as platforms for sharing content, such as music, photographs, and videos. When these websites introduce the ability to create personal profiles, establish contacts, and interactions with other users through comments, they become social networks as well as content hubs. Some popular content-sharing networks include thesixtyone, YouTube, and Flickr.

2.5 **Shared-interest networks**

Some social networks are built around a common interest or geared to a specific group of people. These networks incorporate features from other types of social networks but are slanted toward a subset of individuals, such as those with similar hobbies, educational backgrounds, political affiliations, ethnic backgrounds, religious views, sexual orientations, or other defining interests. Examples of such networks include deviantART, LinkedIn, Black Planet, and Goodreads.

3. **Social networks security requirements**

As people are unaware of the dangers of the sociotechnical attacks, they usually uncover everything about themselves via the internet, thinking that this information does not affect their privacy. The willingness to uncover personal information on SN websites negatively affects their professional and even their personal lives. The security firms Sophos and Nagy and Pecho conducted a study on social network users to test their awareness of protecting personal information. Both studies showed profiles of midrange, college-educated users living in a modern city. Sophos conducted the study in the European region while Nagy and Pecho tested users in the United States. When comparing the results of both studies, it was found that the understanding of what is considered critical information varies between Americans and Europeans. Generally, information about residence and career got lower response from Americans. However, Europeans were more protective of personal information such as phone numbers, emails, Instant Messenger (IM) contacts, and education. In general, there are some basic requirements for security in SN sites, including the following.

Registration and login: The registration phase is needed to grant network access to the user for later authentication. After registering, users acquire a unique User ID and authentication information that should be stored confidentially and with integrity.

Access control: Distinguishing between user and group access privileges is very important to determine the user's and group's ability to access the private and/or shared items or the different profiles. Access control is also needed to manage groups with thousands of users. So access control aims to ensure the integrity, confidentiality, and availability of shared items.

Secure communication: During instant messaging and live chats, messages must be directly sent to the addressed receiver, so senders and receivers must be authenticated, and the communication itself should provide confidentiality and integrity.

4. **SN security technical**

Challenges: SN sites are perfect for illegal online activities, as they consist of a huge number of users with high levels of trust among them. As a result, there is a high range of security risks, threats, and challenges. Here are some of these security challenges and some proposed solutions from the surveyed projects, which will be discussed in the next sections.

4.1 Privacy risks

SN sites provide some mechanisms for privacy settings to protect users, but these mechanisms are not enough to protect the users. The primary privacy problem is that SN sites are not informing users of the dangers of spreading their personal information. Thus, users are not aware of the extent of the risks involved. The second problem is the privacy tools in SN sites, which are not easy to use and do not offer the flexibility for users to customize their privacy policies according to their needs. The third problem is the users themselves cannot control what other users can reveal about them, such as tagging their photos or related information to other friend's profiles.

4.2 Security risks

Social engineering uses fake SN accounts to notify members to reset their accounts or they send malicious e-mails as security notifications. Revealing personal data is another security risk as SN sites may contain many applications developed just for hacker disturbance and revealing user's personal data. Drive-by download is also considered a security risk. Some SN applications may include malicious links or codes, such as requesting users to download a specific program to run these applications. Phishing is another security risk. Some applications such as psychological tests and different type of personality tests use fake web pages just to collect user data through some seemingly harmless questions. For example, in some applications the user is requested to enter the cellular phone number to receive the test result; using this number, the attacker can embezzle some money from the phone's credit. Sometimes phishing attackers send a message to a user's SN posing as a friend asking the user to visit a malicious and fake SN login page. If the user goes to the page, the attacker can steal the user's login credentials to initiate future attacks. In addition, Trojans are considered a security risk as they use fake SN accounts to send emails to users to update their passwords. These usually ask the users to open an attached file to update their password.

Once that is done, a Trojan or other malicious program will be downloaded. A fake friendship invitation is a security risk also that performs identity theft by using false data from a friend to add other friends and to steal their data. Criminal cases can be considered a security risk as terrorists are using SN sites to achieve their targets.

4.3 Anonymity risks

These attacks target the users' identity and privacy. A direct anonymity attack tries to compromise the users' anonymity by exposing their information and location while connecting to an SN site via mobile devices. As mentioned, devices that are in the same system can find the name of the SN user, which will directly compromise the user's privacy. An indirect or K-anonymity attack takes place when several pieces of information that are independently noncompromising indirectly compromise the user's identity by putting these related pieces of information together to map back to the identity of that user. For example, list of favorite stores and restaurants of a certain user can easily be used to track a user and find his/her location.

4.4 Other risks

There are many other sources of risks on SN that may compromise the user and or the SN providers. Some of these may be physical risks, though most are logical. Some examples include the following:

- o Connecting home devices to an SN site may expose more information about the users within the household. This may cause unprotected devices to compromise the others within the home network.
- o Protection from SN site operators is essential since exposure to the operators and perhaps their partners may cause risks on the SN and its users. The possibility of large-scale privacy breaches on the SN sites can compromise some or all user data available on the SN site. An intentional or even unintentional data disclosure will cause a large-scale privacy breach.
- o User dependence on the service provider's existence may also be risky since many of the SN sites are run as free services, and those providers may disappear any time, which basically leaves users with no access to any of the information they keep on the sites. Spoofing the system using false identities or colluding in small groups may provide access to the personal information of others.
- o Trust among users and with service providers is hard to achieve, and the lack of it poses great threats on the success of SN sites. Users need to establish strong trust relationships with each other and with the service providers and in many cases with the service providers' partners. This is a challenge since it is hard to provide concrete methods to create trust or maintain it.

5. What information is public?

There are two kinds of information that can be gathered about a user from a social network: information that is shared and information gathered through electronic tracking.

5.1 Information a user shares

5.1.1 Information a user shares may include the following:

- photos and other media
- age and gender
- biographic information (education, employment history, hometown, etc.).
- status updates (also known as posts)
- contacts
- interests
- geographic location

This information becomes public in a variety of ways:

A user may choose to post information as "public" (without restricting access via available privacy settings).

Certain information may be publicly visible by default. In some situations, a user may be able to change the privacy settings to make the information "private," so only approved users can view it. Other information must remain public; the user does not have an option to restrict access to it.

A social network can change its privacy policy at any time without a user's permission. Content that was posted with restrictive privacy settings may become visible when a privacy policy is altered.

Approved contacts may copy and repost information, including photos, without a user's permission, potentially bypassing privacy settings.

Third-party applications that have been granted access may be able to view information that a user or a user's contacts post privately.

Social networks themselves do not necessarily guarantee the security of the information that has been uploaded to a profile, even when those posts are set to be private. This was demonstrated in one May 2010 incident during which unauthorized users were able to see the private chat logs of their contacts on Facebook. While this and other similar bugs are usually quickly fixed, there is great potential for taking advantage of leaked information.

5.2 Information gathered through electronic tracking

Information may also be gathered from a user's actions online using "cookies" (short strings of text stored on one's hard drive). Some of the purposes of cookies may include these:

- tracking which websites a user has viewed
- storing information associated with specific websites (such as items in a shopping cart)
- tracking movement from one website to another
- building a profile around a user

In fact, a 2009 study conducted by AT&T Labs and Worcester Polytechnic Institute found that the unique identifying code assigned to users by social networks can be matched with behavior tracked by cookies. This means that advertisers and others are able to use information gleaned from social networks to build a profile of a user's life, including linking browsing habits to one's true identity.

5.3 Who can access information?

When posting information to a social network, a user probably expects authorized contacts to be able to view it. But who else can see it, and what exactly is visible?

Entities that collect personal information for legal purposes include these:

- advertisers interested in personal information so they can better target their ads to those most likely to be interested in the product,
- third-party software developers who incorporate information to personalize applications, such as an online games that interact with the social network,

Entities that collect personal information for illegal purposes include these:

- identity thieves who obtain personal information either based on information a user posts or that others post about the user
- other online criminals, such as people planning to scam or harass individuals or infect computers with malware (malicious software placed on a computer without the knowledge of the owner)

6. Fraud on social networks

Criminals may use social networks to connect with potential victims. This section discusses some of the typical scams and devices used to defraud consumers on social networks. Fraud may involve more than one of the techniques described subsequently.

6.1 Identity theft

Identity thieves use an individual's personal information to pretend to be them, often for financial gain. The information users post about themselves on social networks may make it possible for an identity thief to gather enough information to steal an identity. In 2009, researchers at Carnegie University Mellon published a study showing that it is possible to predict most and sometimes all of an individual's 9-digit Social Security number using information gleaned from social networks and online databases.

Information often targeted by identity thieves includes:

- passwords
- bank account information
- credit card numbers
- information stored on a user's computer such as contacts
- access to the user's computer without his or her consent (for example, through malware)
- Social Security numbers. Remember that the key to identity theft is the Social Security number. Never provide a Social Security number through a social networking service.

6.1.1 Some fraud techniques to watch for

Illegitimate third-party applications. These rogue applications may appear similar to other third-party applications but are designed specifically to gather information. This information may be sold to marketers but could also be useful in check identity theft. These applications may appear as games, quizzes, or questionnaires in the format of "What Kind of Famous Person Are You?"

False connection requests. Scammers may create fake accounts on social networks and then solicit others to connect with them. These fake accounts may use the names of real people, including acquaintances, or may be entirely imaginary. Once the connection request is accepted, a scammer may be able to see restricted and private information on a user's profile.

Malware (malicious software) is a term that describes a wide range of programs that install on a user's computer often through the use of trickery. Malware can spread quickly on a social network, infecting the computer of a user and then spreading to his or her contacts. This is because the malware may appear to come from a trusted contact, so users are more likely to click on links and/or download malicious programs.

Some common techniques used in spreading malware include these:

- shortened URLs, particularly on status update networks or newsfeeds; these may lead the user to download a virus or visit a website that will attempt to load malware on a user's computer,
- messages that appear to be from trusted contacts that encourage a user to click on a link, view a video, or download a file,
- an email appearing to be from the social network itself, asking for information or requesting a user click on a link,
- third-party applications that infect computers with malicious software and spread it to contacts,
- fake security alerts, which are applications that pose as virus protection software and inform the user that his or her security software is out of date or a threat has been detected.

7. Tips to stay safe, private, and secure

There are many ways that information on social networks can be used for purposes other than what the user intended. Next are some practical tips to help users minimize the privacy risks when using social networks. Be aware that these tips are not 100% effective. Any time you choose to engage with social networking sites, you are taking certain risks. Common sense, caution, and skepticism are some of the strongest tools you have to protect yourself.

Registering an account:

- Use a strong password different from the passwords you use to access other sites.
- Never provide a work-associated email to a social network, especially when signing up. Consider creating a new email address strictly to connect with your social networking profile(s).
- Consider not using your real name, especially your last name. Be aware that this may violate the terms of service of some social networks.
- Review the privacy policy and terms of service before signing up for an account.
- Be sure to keep strong antivirus and spyware protection on your computer.

- Provide only information that is necessary or that you feel comfortable providing. When in doubt, err on the side of providing less information. Remember, you can always provide more information to a social network, but you cannot always remove information once it is been posted.
- During the registration process, social networks often solicit a new user to provide an email account password so the social network can access the user's email address book. The social network promises to connect the new user with others they may already know on the network. To be safe, do not provide this information at all. There are some social networks that capture all of a user's email contacts and then solicit them—often repeatedly—to join. These messages may even appear to be from the original user. If you consider providing an email address and account password to a social network, read all agreements very carefully before clicking on them.
- Become familiar with the privacy settings available on any social network you use. On Facebook, make sure that your default privacy setting is "Friends Only" Alternatively, use the "Custom" setting and configure the setting to achieve maximum privacy.
- Stay aware of changes to a social network's terms of service and privacy policy. You may be able to keep track of this by connecting to an official site profile, for example, Facebook's Site Governance.
- Be careful when you click on shortened links. Consider using a URL expander (as an application added to your browser or a website you visit) to examine short URLs before clicking on them. Example of URL expanders include LongURL, Clybs URL Expander, and Long URL Please (Privacy Rights Clearinghouse does not endorse one URL expander over another.)
- Be very cautious of pop-up windows, especially any that state your security software is out of date or that security threats and/or viruses have been detected on your computer. Use your task manager to navigate away from these without clicking on them; then run your spyware and virus protection software.
- Delete cookies, including flash cookies, every time you leave a social networking site.
- Remember that whatever goes on a network might eventually be seen by people not in the intended audience. Think about whether you would want a stranger, your mother, or a potential boss to see certain information or pictures. Unless they are glowing, do not post opinions about your company, clients, products, and services. Be especially cautious about photos of you on social networks, even if someone else placed them there. Do not be afraid to untag photos of yourself and ask to have content removed.
- Do not publicize vacation plans, especially the dates you will be traveling. Burglars can use this information to rob your house while you are out of town.
- If you use a location-aware social network, do not make public where your home is because people will know when you are not there. In fact, you should be careful when posting any sort of location or using geotagging features because criminals may use it to secretly track your location. For the same reason, be

careful not to share your daily routine. Posting about walking to work, where you go on your lunch break, or when you head home is risky because it may allow a criminal to track you.

- Be aware that your full birth date, especially the year, may be useful to identity thieves. Do not post it, or at a minimum restrict who has access to it.
- Do not post your address, phone number, or email address on a social network. Remember scam artists as well as marketing companies may be looking for this kind of information. If you do choose to post any portion of this, use privacy settings to restrict it to approved contacts.
- Use caution when using third-party applications. For the highest level of safety and privacy, avoid them completely. If you consider using one, review the privacy policy and terms of service for the application.
- If you receive a request to connect with someone and recognize the name, verify the account holder's identity before accepting the request. Consider calling the individual, sending an email to his or her personal account, or even asking a question only your contact would be able to answer.
- If you receive a connection request from a stranger, the safest thing to do is to reject the request. If you decide to accept the request, use privacy settings to limit what information is viewable to the stranger and be cautious of posting personal information to your account, such as your current location as well as personally identifiable information.
- Be wary of requests for money, even if they are from contacts you know and trust. If a contact's account is compromised, a scam artist may use his or her name and account to attempt to defraud others through bogus money requests.
- Take additional precautions if you are the victim of stalking, harassment, or domestic violence.
- In the event that your social networking account is compromised, report it to the site immediately and alert your contacts. You will need to change passwords but proceed with caution because your computer security may have been compromised. Malware, including key-logging software, may have been installed on your computer. If you use online banking, do not log on from the computer that may have been compromised until you have ensured your computer security is intact.
- Prune your "friends" list on a regular basis. It is easy to forget who you have friended over time, and therefore with whom you are sharing information.
- If you are using a social networking site that offers video chatting, pay attention to the light on your computer that indicates whether or not your webcam is in use. This will help you avoid being "caught on camera" by accident.
- Be sure to log off from social networking sites when you no longer need to be connected. This may reduce the amount of tracking of your web surfing and will help prevent strangers from infiltrating your account.
- Remember that nothing that you post online is temporary. Anything you post can be cached, stored, or copied and can follow you forever. Check your privacy settings often. Privacy policies and default settings may change, particularly on Facebook.

8. Staying safe on social media sites

Each social media site offers tips on how to use their service and still maintain a high level of security. Read their policies, follow their security guidelines, and adopt their best practices.

- **Facebook:** There is a comprehensive help page on Facebook where you can find details on protecting your account against hacking and other security threats. Check it frequently to make sure your practices and settings are up to date. CNET also offers practical advice such as being sure to block your ex and carefully manage who has viewing access.
- **Foursquare:** For a better understanding of who can see information associated with your Foursquare account, visit the Help Center. This page explains methods for creating security settings for every account scenario.
- **Instagram:** If you have an Instagram account, read their official page for ways to keep your account safe.
- **LinkedIn:** Visit LinkedIn's Help Center for a wide range of account security articles. A few of the topics covered on the page include methods for protecting your privacy, your identity, and your account. They also offer tips for dealing with phishing, spam, and malware. If your LinkedIn account is associated with a business, How Not to Have Your Account Hacked provides ways to keep passwords safe even if several people have access to the account.
- **Pinterest:** To keep your Pinterest account secure, you will need to access two main sections on the site: privacy settings and account security. If your account has been hacked or placed in a safe mode by Pinterest, you will use the account security section to resolve the issue. However, most likely you will only need to use the privacy settings section. This is where you control what others can view and the degree of personalization desired. Scams are one issue the site has dealt with in the past.
- **Tumblr:** If you use Tumblr, one of the best ways to improve security is to utilize the recently implemented two-factor authentication. For all your settings, though, access the site's security settings page. Here you can learn how to revoke third-party application permissions as well as how to remove spam from your blog. For increased security, according to Entrepreneur magazine, you may want to refrain from using free themes.
- **Twitter**: Visit Twitter's Help Center to learn best practices for your Tweets or if you want to know how to connect with or revoke third-party applications. Also visit this page to discover methods for controlling account settings so you can get the level of security you want.
- **YouTube/Google+:** If you have a YouTube and/or Google+ account, bookmark Google's Keeping Your Account Secure page. This page is great source to learn about their two-step verification process, malware and virus issues, general information about your account settings, and best practices for protecting your privacy and identity.

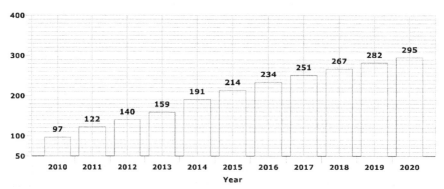

FIGURE 1.1

Number of SNS users worldwide from 2010 to 2016 with prediction to 2020 (Statista's report).

One situation people sometimes overlook is what to do if they want to close a social media account. Should the account be deactivated or deleted? According to the Center for Internet Security, you need to take several steps before your account is deleted from the social media site.

The Legal Technology Center cites two federal statutes governing social networking sites and legal issues in 2019 (Fig. 1.1).

9. SNS security issues and challenges

- Sophos' security threat report 2011 reveals that Facebook has the biggest security risks, which are significantly ahead of MySpace, Twitter, and LinkedIn.
- Facebook is the most popular site for active users on the web.
- Due to this popularity, a large amount of users are targeted by adversaries via various types of attacks, such as malware, phishing, spamming, and more (Fig. 1.2)
- The Kaspersky Security Network has described a parental control component that supports parents in precaution their kids from the concealed risks of abandoned use of computers and the internet.
- A worldwide analysis of this component with various real-world security risks demonstrates that it is prompted most often by social network risks.
- This indicates that SNSs act as an escalating noticeable role in kids' lives, and parents are increasingly worried that their children are more vulnerable to security risks with the use of SNSs (Fig. 1.3).
- According to the internet security threat report, SNSs have become the favorite target of scammers in the past few years.
- They use various scamming techniques to scam SNS users via the usage of manual sharing; fake offerings, like jacking; fake applications; and fake plugins.
- However, manual sharing has been used more widely in recent years. Fig. 1.4 shows the percentage utilization of each scamming technique for the last 3 years

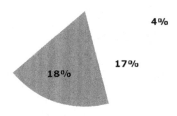

FIGURE 1.2

Sophos security threat report, 2011.

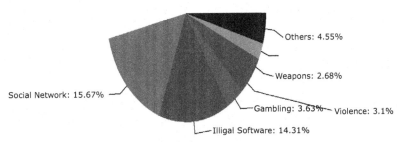

FIGURE 1.3

Analysis of parental control component triggered by various real-world security risks.

10. Five top social media security threats

Social media platforms such as Twitter, Facebook, and LinkedIn are being used by enterprises to engage with customers, build their brands, and communicate information to the rest of the world.

But social media for enterprises is not all about "liking," "friending," "up-voting" or "digging." For organizations, there are real risks to using social media, ranging from damaging the brand to exposing proprietary information to inviting lawsuits.

Here are five of the biggest social media security threats:

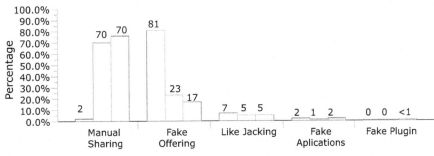

FIGURE 1.4

The percentage utilization of various SNSs scamming techniques for the last 3 years.

10.1 Mobile apps

The rise of social media is inextricably linked with the revolution in mobile computing, which has spawned a huge industry in mobile application development. Naturally, whether using their own or company-issued mobile devices, employees typically download dozens of apps because, well, because they can. But sometimes they download more than they bargained for. In early March, Google removed from its Android Market more than 60 applications carrying malicious software. Some of the malware was designed to reveal the user's private information to a third party, replicate itself on other devices, destroy user data, or even impersonate the device owner.

10.2 Social engineering

A favorite of smooth-talking scammers everywhere, social engineering has been around since before computer networks. But the rise of the Internet made it easier for grifters and flimflam artists to find potential victims who may have a soft spot in their hearts for Nigerian royalty.

Social media has taken this threat to a new level for two reasons: (1) people are more willing than ever to share personal information about themselves online via Facebook, Twitter, Foursquare, and MySpace, and (2) social media platforms encourage a dangerous level of assumed trust. From there, it is a short step to telling your new friend about your company's secret project. And your new friend really might be able to help with this if you would only give him a password to gain access to a protected file on your corporate network. Just this once.

10.3 Social networking sites

Sometimes hackers go right to the source, injecting malicious code into a social networking site, including inside advertisements and via third-party apps. On Twitter, shortened URLs (popular due to the 140-character tweet limit) can be used to trick users into visiting malicious sites that can extract personal (and corporate) information if accessed through a work computer. Twitter is especially vulnerable to this method because it is easy to retweet a post, so it eventually could be seen by hundreds of thousands of people.

10.4 Your employees

You knew this was coming, but even the most responsible employees have lapses in judgment, make mistakes, or behave emotionally. Nobody is perfect all of the time.

But dealing with an indiscreet comment in the office is one thing; if the comment is made on a work-related social media account, then it is out there, and it cannot be retrieved. Just ask Ketchum PR Vice President James Andrews, who 2 years ago fired off an infamous tweet trashing the city of Memphis, hometown of a little

Ketchum client called FedEx, the day before he was to make a presentation to more than 150 FedEx employees (on digital media, no less!).

The tweet was discovered by FedEx employees, who emailed angry missives to Ketchum headquarters protesting the slight and wondering why FedEx was spending money on a snooty New York PR firm while employees were dealing with a 5% salary cut during a severe recession. Andrews had to make a very public and humiliating apology.

10.5 Lack of a social media policy

This one is totally on you. Without a social media policy for your enterprise, you are inviting disaster. You cannot just turn employees loose on social networking platforms and urge them to "represent." You need to spell out the goals and parameters of your enterprise's social media initiative. Otherwise you will get exactly what you are inviting: chaos and problems.

Who is allowed to use social media on behalf of the organization and what they are allowed to say are the two questions that must be most clearly addressed in a social media policy. You need to make all this clear or employees will make decisions on their own, on the fly. Does that sound like a good thing?

Two more imperatives related to social media policy are these: (1) organizations must conduct proper training for employees, if only to clear up issues regarding official social media policies, and (2) a social media initiative needs a coordinator and champion, i.e., a social media manager.

11. Privacy and security of social networks for home users

In the times in which more and more people have access to the Internet and the Internet has become an exchange platform of information, it is playing an increasingly greater role in everyday life, and the importance of privacy and security of social networks on these platforms to maintain social connections such as Facebook, XING, Google+, and many more is growing.

These social networks are enjoying a growing number of members every day, and users are often used to remaining always and everywhere connected with friends and acquaintances to exchange views on a variety of things. For example, Facebook sign in is approximately 50% of all the users every day at least once on average, and it calls more than 350 million users.

These platforms are designed for the end user to use for free, but the providers must cover their costs somehow and make their money in their part by selling data or the display of personalized and targeted advertising. The amount of data that each user inserts to the network daily, of course, should also be protected. Careless handling of personal information, even after a long time, might lead to unpleasant surprises.

12. Social media security tips to mitigate risks

With the amount of information stored and shared online, social media security is more important than ever.

There is no doubt social media has made the world a more connected place. In most cases, that is a good thing. But all those connections also create unprecedented access to people's and business' information. And that can be a very bad thing when hackers and scammers get involved.

Giving up social media is not a reasonable option. But neither is it reasonable to carry on as if social networks are always safe and secure. You need to take steps to protect your company against some of the most common social media security threats. Here is where to start.

12.1 Common social media security risks

It can be a good idea to reserve your brand's handle on all social media channels, even if you do not plan to use them all right away. This allows you to maintain a consistent presence across networks, making it easy for people to find you.

But it is important not to ignore the accounts you do not use yet, those that you have stopped using, or those you do not use often.

Idle social accounts can be the target of hackers, who could start posting fraudulent messages under your name.

Knowing the account is unmonitored, once they gain control, they could send anything from false information that is damaging to your business to virus-infected links that cause serious problems for followers. And you will not even notice until your customers start coming to you for help.

12.1.1 Human error

Everyone makes mistakes. In today's busy world, it is all too easy for an employee to accidently expose the company to threats online. In fact, 77% of respondents to the 20th EY Global Information Security Survey said that a careless member of staff was the most likely source of a cyber security threat.

12.1.2 Third-party apps

Even if you have your own social accounts on lockdown, hackers may be able to gain access through vulnerabilities in third-party apps that integrate with the big social networks. For example, hackers gained access to the Twitter accounts of Forbes and Amnesty International using a flaw in the Twitter Counter app, used for Twitter analysis.

12.1.3 Phishing attacks and scams

Phishing scams use social media to trick people into handing over personal information (like banking details, passwords, or business information). A recent social media scam involved false reports that the actor Rowan Atkinson had died. (The Mr. Bean and Blackadder actor is still very much alive.) What looked like a video link actually directed users to a page that said their computer had been locked, with a phone number to call for support. Rather than a support team, the phone line connected to scammers looking for credit card numbers and personal information. Worse, the "support software" offered was actually a virus. Filmmaker Tyler Perry recently posted a video to his Facebook account warning fans not to fall victim to giveaway scams using his name. He said his team has to shut down, "10, 20, 30 of these things" every day.

12.1.4 Imposter accounts
The number of fraudulent social support accounts doubled between the third quarter of 2016 and Q3 2017. Those accounts can target your customers, tricking them into handing over confidential information and tarnishing your reputation in the process. Imposter accounts may also try to con your employees into handing over login credentials for corporate systems.

12.1.5 Malware attacks and hacks
Social media hackers are becoming more sophisticated. Hackers have gained access to big-name Twitter accounts, from Kylie Jenner and Mark Zuckerberg, to several HBO shows. Those hacks were fairly benign. Others are much more serious. For example, hackers used a fake profile to connect with employees of targeted organizations, sharing a file that gave the attackers remote access to the victims' work computers.

12.1.6 Privacy settings
People seem to be well aware of the potential privacy risks of using social media. A recent survey found that nearly two-thirds of people have "very little" or "no" trust in social networks when it comes to privacy protection.

12.1.7 Unsecured mobile phones
Mobile devices are the most popular way to access social networks. Using each network's mobile app makes it easy to access social media accounts with just one tap. That is great as long as your phone stays in your possession. But if your phone, or an employee's phone, is lost or stolen, one-tap access makes it easy for a thief to access social accounts. And then they can message all of your connections with phishing or malware attacks. Protecting the device with a password helps, but more than half of mobile phone users leave their phones unlocked.

12.2 Social media security tips and best practices
12.2.1 Create a social media policy
If your business is using social media—or getting ready to—you need a social media policy. These guidelines outline how your business and its employees should use

social media responsibly. This will not only protect you from security threats, but bad PR or legal trouble as well.

At minimum, your social media policy should include the following:

- brand guidelines that explain how to talk about your company on social media
- rules related to confidentiality and personal social media use
- which departments or team members are responsible for each social media account
- guidelines related to copyright and confidentiality
- guidelines on how to create an effective password and how often to change passwords
- expectations for keeping software and devices updated
- how to identify and avoid scams, attacks, and other security threats
- who to notify and how to respond if a social media security concern arises

12.2.2 Train your staff on social media security best practices

Even the best social media policy will not protect your organization if your employees do not follow it. While your policy should be easy to understand, training will give employees the chance to engage, ask questions, and get a sense of how important it is to follow. These training sessions are also an opportunity to review the latest threats on social media and talk about whether there are any sections of the policy that need updating. And it is not all doom and gloom. Social media training also equips your team to use the tools effectively. When employees understand best practices, they will feel confident using social media for both personal and professional purposes.

12.2.3 Limit social media access

While you may be focused on threats coming from outside your organization, PricewaterhouseCoopers found employees are more likely to cause cyber security incidents than are hackers. Limiting access to your social accounts is the best way to keep them secure. You may have whole teams of people working on social media messaging, post creation, or customer service. But that does not mean everyone needs the ability to post. And it does not mean that everyone needs to know the passwords to your social accounts.

The first line of defense is to limit the number of people who can post on your accounts. Think carefully about who needs posting ability and why. Once you have decided who can post, use software like Hootsuite to give the right people the right account access. This way, they never need to know the individual login information for any social network account. If the person leaves your company, you can disable their account without having to change all the social network passwords.

12.2.4 Set up a system of approvals for social posts

Z-Burger recently faced a major crisis after a marketing contractor used a photo of a slain journalist in an extremely inappropriate Twitter post. No one at Z-Burger saw the tweet before it was posted, since they had given the contractor the ability to publish directly to their account.

The owner of Z-Burger was horrified when he saw the tweet and took action to delete the offensive post right away. But if he had set up an approval system, he or his staff would have reviewed the tweet before it was published. And the crisis would have been averted.

You can use Hootsuite to give employees or contractors the ability to draft messages, preparing them so they are all set to post at the press of a button. But leave that last button press to a trusted person on your team.

12.2.5 Put someone in charge

Designating a key person as the eyes and ears of your social presence can go a long way toward mitigating risks. This person should own your social media policy, monitor your brand's social presence, and determine who has publishing access. This person should also be a key player in the development of your social media marketing strategy.

This person will likely be a senior person on your marketing team. But they should maintain a good relationship with your company's IT department to ensure marketing and IT work together to mitigate risk.

This person is also whom team members should turn to if they ever make a mistake on social media that might expose the company to risk of any kind, from security to a damaged reputation. This way the company can initiate the appropriate response.

12.2.6 Monitor your accounts and engage in social listening

As mentioned at the start, unattended social accounts are ripe for hacking. Keep an eye on all of your social channels, from the ones you use every day to the ones you have registered but never used at all. Assign someone to check that all of the posts on your accounts are legitimate. Cross-referencing your posts against your content calendar is a great place to start. Follow up on anything unexpected. Even if a post seems legitimate, it is worth digging into if it strays from your content plan. It may be simple human error. Or, it may be a sign that someone has gained access to your accounts and is testing the water before posting something more malicious. You also need to watch for imposter accounts, inappropriate mentions of your brand by employees (or anyone else associated with the company), and negative conversations about your brand.

12.2.7 Invest in security technology

No matter how close an eye you keep on your social channels, you cannot monitor them 24 h a day, but software can. Solutions like ZeroFOX will automatically alert you of security risks. When you integrate ZeroFOX with your Hootsuite dashboard, it will alert you to dangerous, threatening, or offensive content targeting of your brand; malicious links posted on your social media accounts; scams targeting your business and customers; and fraudulent accounts impersonating your brand. It also helps to protect against hacking and phishing attacks.

12.2.8 Perform a regular audit

Social media security threats are constantly changing. Hackers are always coming up with new strategies, and new scams and viruses can emerge at any time. Scheduling regular audits of your social media security measures will help keep you ahead of the bad actors.

At least once a quarter, be sure to review the following:

- Social network privacy settings. Social media companies routinely update their privacy settings, which can have an impact on your account. For example, a social network might update its privacy settings to give you more precise control over how your data is used.
- Access and publishing privileges. Perform a scan of who has access to and publishing rights on your social media management platform and social media accounts and update as needed. Make sure all former employees have had their access revoked, and check for anyone who has changed roles and no longer needs the same level of access.
- Recent social media security threats. Maintain a good relationship with your company's IT team so they can keep you informed of any new social media security risks of which they become aware. And keep an eye on the news: big hacks and major new threats will be reported in mainstream news outlets.
- Your social media policy. This policy should evolve over time as new networks gain popularity, security best practices change, and new threats emerge. A quarterly review will make sure this document remains useful and helps to keep your social accounts safe.
- Use Hootsuite to manage all your social media accounts safely and securely in one place. Mitigate risks and stay compliant with best-in-class security features, apps, and integrations.

13. Securing Social Security's future

With all the doom and gloom that swirls around the future of Social Security, it is no wonder so many people have questions about what is in store. Only 10% of Americans ages 25 to 69 are "very confident" they will get as much as the program delivers today, and 18% believe they will get nothing, the Employee Benefit Research Institute (EBRI) 2015 Retirement Confidence Survey found.

Yet despite the program's challenges, concerns about its long-term viability go way too far. Most experts agree that Social Security will endure as the foundation of retirement security for the majority of Americans, and not just for boomers but for their children, grandchildren, and great-grandchildren as well.

---

14. Humans want Social Security

Majorities of the public benefit from Social Security—or will 1 day—so pretty much everyone wants to maintain its protections. In a 2014 survey by the National Academy of Social Insurance, 84% of Democrats and 69% of Republicans agreed it is "critical" to preserve benefits, even if that meant increasing taxes. Surveys by CBS News, the Pew Research Center, and the National Institute on Retirement Security have found that Social Security is supported by significant majorities of the public, approaching 70% or higher.

Even young adults, the first to say "it will not be there for me," support the program. While more than one in four millennials do not expect to get any benefits from Social Security, according to the EBRI survey, a Pew survey found last year that 61% opposed cuts.

Political leaders are aware of the poll numbers, which is why it is a tricky topic for anyone to tackle. Experts say this robust public support is another reason to believe our elected leaders ultimately will engage in a debate over keeping the program strong for the long term.

14.1 Most humans need Social Security

In many ways, the program is becoming more important in today's economy, as growing numbers of Americans face retirement without other financial resources to rely on.

Social Security provides at least half the family income for about one in two older Americans, and it provides most of the income for one in four. In part, that is because other financial supports for retirement have weakened in recent years.

Private employer-paid pensions, which pay a set amount for life, are increasingly rare. As one measure, the number of single-employer defined-benefit pension plans protected by the Pension Benefit Guaranty Corp. plunged from a peak of more than 112,000 in the mid-1980s to about 23,000 in 2013.

And nothing has come along to truly replace them. Plans such as 401(k)s and individual retirement accounts have proved inadequate for most people. Counting part-time workers, only about half the labor force even has access to a pension or retirement savings plan through an employer. Of those with access to retirement savings plans, many either fail to participate fully or spend the money before retirement.

For these and other reasons, it is no surprise that many families have little saved for old age.

Almost half of working-age households have nothing saved in a retirement account, according to the National Institute on Retirement Security. One recent survey found that about four in 10 older workers have less than $25,000 in retirement savings.

Women depend on Social security most of all. They have fewer financial resources than men on average, yet they live longer.

Women also are more likely to be single, and single adults often have greater financial need than married couples. One study found that unmarried boomers were almost five times more likely to be poor than married boomers. And the likelihood of being single and older is not just a boomer phenomenon. Millennials are far more likely to be single than were the boomers or Gen Xers at the same stage of their lives.

Taken together, these trends are telling a story: the need for income in retirement is growing, people will rely more on Social Security every month, and they will depend on its benefits longer than ever.

15. Social Security can be strengthened

Doubts about Social Security's future are based on an assumption that the financial challenge the program faces is overwhelming. How big is the long-term shortfall?

According to the system's 2014 Trustees Report (the 2015 report is expected shortly), Social Security has a $2.8 trillion surplus and can pay full benefits for another 18 years. The surplus will be gone in 2033, estimated the report, which forecast the outlook 75 years into the future. At that time, if Congress takes no action, revenue from payroll taxes would enable Social Security to pay people 77% of their promised benefits.

Looked at this way, for the rest of this century, Social Security is about three-fourths stable and one-fourth out of balance. And from a financial standpoint, that is not overwhelming. Experts of all political stripes have suggested dozens of ideas over the years to keep Social Security strong far into the future.

Most fix-it measures entail either increased contributions to Social Security, some modification of benefits for some or all future beneficiaries, or some combination of both. Political leaders will have to work through tough decisions, particularly over who will bear the burden of any changes.

As one often-mentioned example, consider the current wage cap of $118,500, above which no one pays payroll taxes for Social Security. Raising that cap to about $255,000 could solve a little more than one-fourth of the long-range problem, while having no effect on most workers. Advocates of this measure say it would enable Social Security to recapture the share of earnings it received in the 1980s, before a surge of income gains at the top of the ladder reduced its share. Critics say it would harm the economy and affect many workers who are not wealthy.

Other options that may be debated one day include tightening benefits in the future for the more affluent, slowly raising the payroll tax by 1 percentage point or slightly more over a period of many years, and bringing uncovered workers in state and local government into the Social Security system. Some also may call for raising the retirement age to reflect increased longevity, although opponents point out that gains in life expectancy have mostly gone to the affluent and educated.

Those are just examples from a longer list, and each has variations. Further, changes in benefits do not necessarily mean reductions. A growing number of voices are calling for enhanced benefits, particularly for low-income workers and the oldest retirees, whose Social Security payments may not even lift them above the poverty line.

16. Conclusion

Social network sites are a major application driver with millions of users all over the world relying on them for keeping contact and sharing information with others. This huge involvement drives the need for setting the right security measures that help in protecting user privacy. In this paper, we discussed a number of mechanisms and approaches that help in achieving acceptable levels of security for the SN providers and users. However, many of these mechanisms provided a solution for a certain privacy concern but missed others. Moreover, when it comes to setting higher security measures, they seem to compromise the usability and flexibility of the system for the average user. However, all surveyed projects failed to mention or measure the tradeoffs between higher security measures and the systems' performance. There are many opportunities for new mechanisms or even existing mechanisms to explore these areas to try to design mechanisms that will not require (or at least minimize) tradeoffs when it comes to users' privacy, information security, usability, flexibility, and performance.

The increased public attention to the handling of data and collection from the social networks has reworked a variety of configuration options for privacy. However, this affects only the other users, whether they can see the information or not. The providers of social networks have complete access to the complete data released about a user.

A simple agreement to terms and conditions of usage may lead to unwanted surprises if one suddenly finds his/her data at an unexpected location. In addition, anyone must be aware that in a social network the provider can delete a user's account, but it is difficult if not impossible to erase all the data, since these are usually already redistributed. As they say, "the Internet never forgets."

References

[1] F. Al-Turjman, Intelligence and security in big 5G-oriented IoNT: an overview, Elsevier Future Generat. Comput. Syst. 102 (1) (2020) 357−368.

[2] Y. Qadri, R. Ali, A. Musaddiq, F. Al-Turjman, D. Kim, S. Kim, The Limitations in the State-of-the-Art Counter-measures against the Security Threats in H-IoT, Cluster Computing, 2020, https://doi.org/10.1007/s10586-019-03036-7.

Further reading

[1] S. Jabbar, S. Khalid, M. Latif, F. Al-Turjman, L. Mostarda, Cyber security threats detection in internet of things using deep learning approach, IEEE Access 7 (1) (2019) 124379−124389.

[2] J. Wang, S. Jabbar, F. Al-Turjman, M. Alazab, Source code authorship attribution using hybrid approach of program dependence graph and deep learning model, IEEE Access 7 (1) (2019) 141987−141999.

[3] D. Gritzalis, M. Kandias, V. Stavrou, L. Mitrou, History of information: the case of privacy and security in social media, in: Proc. of the History of Information Conference, 2014, pp. 283−310.

[4] S. Deliri, M. Albanese, Security and privacy issues in social networks, in: Data Management in Pervasive Systems, Springer, Cham, 2015, pp. 195−209.

[5] B. Krishnamurthy, Privacy and online social networks: can colorless green ideas sleep furiously? IEEE Secur. Priv. 11 (3) (2013) 14−20.

[6] N.F. Othman, R. Ahmad, M. Yusoff, Information security and privacy awareness in online social networks among UTeM undergraduate students, J. Hum. Capital Dev. 6 (1) (2013) 101−110.

[7] B. Carminati, E. Ferrari, Privacy-aware collaborative access control in web-based social networks, in: IFIP Annual Conference on Data and Applications Security and Privacy, Springer, Berlin, Heidelberg, July, 2008, pp. 81−96.

[8] S. Ali, N. Islam, A. Rauf, I.U. Din, M. Guizani, J.J. Rodrigues, Privacy and security issues in online social networks, Future Internet 10 (12) (2018) 114.

[9] N. Senthil Kumar, K. Saravanakumar, K. Deepa, On privacy and security in social media−a comprehensive study, Procedia Comput. Sci. 78 (2016) 114−119.

[10] R. Gross, A. Acquisti, Information revelation and privacy in online social networks, in: Proceedings of the 2005 ACM Workshop on Privacy in the Electronic Society, November, 2005, pp. 71−80.

[11] L.A. Cutillo, M. Manulis, T. Strufe, Security and privacy in online social networks, in: Handbook of Social Network Technologies and Applications, Springer, Boston, MA, 2010, pp. 497−522.

[12] M. Xue, Y. Liu, K. Ross, H. Qian, Thwarting location privacy protection in location-based social discovery services, Secur. Commun. Network. 9 (11) (2016) 1496−1508.

[13] Y. Gao, N. Zhang, Social Security and Privacy for Social IoT Polymorphic Value Set: A Solution to Inference Attacks on Social Networks, Security and Communication Networks, 2019, pp. 1−16, 2019.

[14] G.J. Ahn, M. Shehab, A. Squicciarini, Security and privacy in social networks, IEEE Internet Comput. 15 (3) (2011) 10−12.

[15] V. Arnaboldi, A. Guazzini, A. Passarella, Egocentric online social networks: analysis of key features and prediction of tie strength in Facebook, Comput. Commun. 36 (10−11) (2013) 1130−1144.

[16] P.C. Pimenta, C.M. de Freitas, June). Security and privacy analysis in social network services, in: 5th Iberian Conference on Information Systems and Technologies, IEEE, 2010, pp. 1−6.

[17] F. Al-Turjman, H. Zahmatkesh, An Overview of Security and Privacy in Smart Cities' IoT Communications, Wiley Transactions on Emerging Telecommunications Technologies, 2019, https://doi.org/10.1002/ett.3677.

Emerging social information networks applications and architectures

Dasari Siva Krishna[1], Patruni Muralidhara Rao[2], Thammada Srinivasa Rao[3]

[1]*Computer Science and Engineering, GMR Institute of Technology, Srikakulam, Andhra Pradesh, India;* [2]*School of Computer Science and Engineering, Vellore Institute of Technology, Vellore, Tamil Nadu, India;* [3]*Computer Science and Engineering, GITAM Institute of Technology, GITAM, Visakhapatnam, Andhra Pradesh, India*

1. Introduction

1.1 Introduction to web: historical aspects of the web relative to the social network

In the 1990s, Web 1.0 was be the first stage of development of the World Wide Web (WWW) that pertained to a simple static web page. Those days, websites were designed as static pages, and few were embedded with HTML markup [1]. The key challenges in the 1990s were customers/users interacting with the website along with the features that are plugged into the browsers and marked in the development of the WWW. In the later stage of Web 1.0, web development did business in a wide range of expenses both in a technologic point view and user experience. Some of the key characteristics are content, search, commerce, static page, combination of content and layout, and emailing of forms [2]. In the transformation from Web 1.0 to 2.0, servers were plugged and upgraded. The data growth in average connection speeds rapidly increased and was vibrant to new technologies for developers. Some of the key characteristics were speed, collaboration, trustworthiness, and improved interoperability. In the present context, Web 3.0 refers to Semantic Web and applications of artificial intelligence [3]. Web relevance is most preciously given to semantic information, and these elements are semantically connected in the social network and its application.

Some of the key characteristics are that it is ubiquitous, individualized, and efficient (Fig. 2.1).

Today, social networking can exponentially grow as a communication media that permits people to share their personal and professional information, create content, and have conversations among users. Consequently, social networking sites can be widely used to share billions of people's personal information, informing themselves, across the world. Moreover, the users in social media have their own sites,

Security in IoT Social Networks. https://doi.org/10.1016/B978-0-12-821599-9.00002-9
Copyright © 2021 Elsevier Inc. All rights reserved.

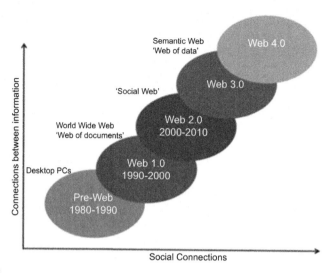

FIGURE 2.1

Evolution of the web.

namely blogs, social network accounts, photo sharing, video sharing, etc., in the context of business organizations. To get feedback from various customers, they have a conversation with their target audience, to advertise and increase their branding. Moreover, authorities also use social media to increase their knowledge of their respective governances and build a network of professionals from related industries.

1.2 Introduction to social networks

In recent years, SNs have been emerging in supporting many novel applications. These provide a lot of attention toward application development and related services. Compared to traditional social networks, these are centered and have some additional capabilities because of the current usage of mobile devices including smartphones. They have several capabilities for devices including global position system (GPS), sensing modules, and radios. This enables SNs to increase traditional social networks with advanced features including location awareness [4]. Fig. 2.2 depicts a three-level architecture. The first layer deals with servers that provide databases, network services, and processing elements. The second layer deals with wireless access networks that handle access points and gateways. The third layer deals with client-side devices like smartphones, which contain smart components, terminals, and GPS architectures.

There are many social applications currently used. Each application has features to give a better experience for the user in the social background. Table 2.1 shows some of the applications and the key features.

In present new technology, many applications are developed based on the latest tool. Table 2.2 shows the frontend and backend technology used for the specific feature.

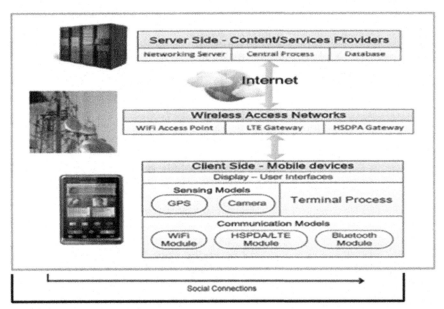

FIGURE 2.2

Social network elements.

Table 2.1 Social application features.

Name of the application	Features
Facebook	- Retrieve and post data to facebook - Connect users and share information - Facebook-based feedback system
Twitter	- Tweet content of the website - Display tweet with expanded media (e.g., photo, video, etc.)
Google+	- Connect with others for better engagement - Use Google + features
YouTube	- Search and view videos - Retrieve standard feeds - Authenticate users to upload videos
Amazon	- Cloud services - Customer services - Covers a vast elastics resources at the infrastructure layer - Sufficient SDKs for mobile operating systems

Table 2.2 Social networking development tools.

Feature	Frontend	Backend
Login	Facebook API, google API, ReactJS	*Django -all -auth, Django -rest -auth, Auth0, Django -rest -framework -jwt, PostgreSQL /Redis for sessions*
User profiles	Location picker cropit, ReactJS	Custom logic on top of django and/or DRF, PostgreSQL for data storage
Connections	React infinitely scroller, ReactJS	Custom logic on top of django and/or DRF, PostgreSQL/Redis for data storage
Messaging	WebSockets, ReactJS	WebSockets through tornado/aiohttp + redis, centrifugo, PostgreSQL/RabbitMQ for message storage and queuing
Creating posts	HTML5, ReactJS	DRF and DRF serializers for HTTP and logic han/Algolia
Uploading photos	HTML5, react Player, ReactJS	For small projects, easy thumbnails; otherwise thumbor for handling thumbnails
Push notifications	React toastr, ReactJS	PushWoosh, firebase cloud messaging
Feed	React infinitely scroller, react - play, ReactJS	Custom logic on top of django /DRF and PostgreSQL /Redis for storage, caching, and pagination

1.3 Social networks architecture

A social network is a place that expands information, where a number of one's business and/or social contacts make connections through individual elements. The elements or resources in the network are growing rapidly in the present day. In the context of social media sites such as Facebook, Twitter, LinkedIn, and Google+ the information is shared among individuals. This creates more exposure to social network architectures and technology like big data, the cloud, blockchain, etc. Fig. 2.3 shows the traditional model of a social network with a layered view of social network architecture, where the first layer is relative to data models that have data representation, business logics, and data services. The second layer is integration between the client tool models and server applications. In the third layer the base components for an application that depends on deployment services concerning their applications are shown [5].

Social network platforms accomplish socialization for a wide range of users. Moreover, these are users of rapidly increasing applications like Facebook, Twitter, and LinkedIn. Similar to the increasing developments of the SNs, they brought up the terms of social media. It is a channel where users communicate and interact with each other on the internet. However, it formed by the collection of contents

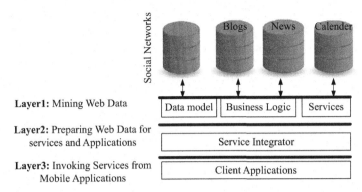

FIGURE 2.3

Layered view of a social network architecture.

that are produced by users in a frame; Fig. 2.3 shows the social media applications that are used in SNs.

In social network applications like Facebook and WhatsApp, there are multiple data centers. The network traffic carries the information from server to server and user to servers. In past years, these have been connected by a single wide-area backbone network known as a WAN [6]. Nowadays, the bandwidth demand for the cross-data center reproduces content like text, images, audio, and video data. In this context, the traffic from the application has increased rapidly. The key challenge in the classic backbone network is to control the traffic, speed of access, and efficiency in storage (Fig. 2.4).

Further, machine-to-machine traffic has occurred in large bursts, interfering with regular user traffic. As data centers increase, the network traffic has split the internet-facing traffic to the cross-data center. The key issues are optimizing the different networks individually.

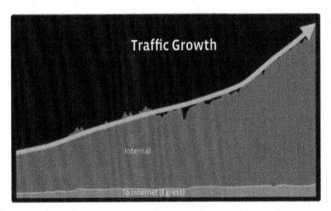

FIGURE 2.4

Network growth relative to the users.

The Express Backbone (EBB) was introduced as a new cross-data center backbone network since network traffic has grown rapidly.

Of late, recommended social network systems create pros and cons for the user experience.

These systems target social media domains and attract attention to the growing popularity of online social networks (OSNs) [7]. The significant OSNs are User Interest Modeling (UIM) and Recommendation Algorithm. In recent studies, DBpedia has used a user as a resource in the Semantic Interest Graph (SIG), which shows an efficient UIMs in OSNs. The key challenge in SIG can be the similarity between the resources (Tables 2.3–2.5).

2. Background with new technologies

2.1 Introduction to Semantic Web technology

In the 20th century, we have witnessed a new revolution in computer and communication systems. The internet is the fastest communication system for sharing information, attracting 50 million people within a short span of 4 years from its inception. Other communication channels like radio and television took 38 and 13 years, respectively, to attract the same number of people. The technology changed concerning the growth of data as well as problems that are solved by researchers. In the present version of the web, social networks are popular in dynamically generating data. This creates the technology changes from Web 2.0 to Web 3.0, i.e., Semantic Web, that provide machine-readable translations [8].

In the new technology, Semantic Web, named by Tim Berners-Lee, is processed by machines. It is a collaborative movement headed by the international standards body World Wide Web Consortium (W3C), providing automatic machine-readable common data formats. In this context, the Semantic Web majorly provides

Table 2.3 Important notations used.

Acronym	Description	Acronym	Description
NLP	Natural language processor	RDF	Resource description framework
AI	Artificial intelligence	SW	Semantic Web
KR	Knowledge representation	XML	Extensible Markup Language
DL	Description logic	URI	Uniform resource identifier
SNs	Social networks	IRI	Internationalized resource identifiers
SNSs	Social network sites	OWL	Web Ontology Language
W3C	World Wide Web consortium	SWRL	Semantic Web Rule Language

Table 2.4 Comparison of planning domain modeling languages.

Name of planning languages	Application	Features
DDL (planning domain definition language)	Robot path planning	Preconditions, actions, effects
NDDL (new domain definition language)	Control of autonomous vehicle robot systems	Timelines, activities, and capabilities
DDL (domain definition language)	Robot path planning	State variables, state variable values, compatibilities
ANML (Action notation modeling Language)	Robot task planning	Defined semantics, variable value, function representation, and rich temporal constraints
AML (ASPEN modeling Language)	ASPEN framework supports a variety of planning and scheduling applications (spacecraft, autonomous robot)	Action, state, resource, temporal constraints, preplanning, during, and plan execution

Table 2.5 Comparison of ontology editors.

Name of tools	Language	Application
Protégé	RDF	Robot path planning
OILed	DAML + OIL	Autonomous vehicle path planning

a semantic framework that allows a user's data to be shared and reused. However, this encourages the inclusion of semantic content to the present web of data. The Semantic Web protocol stack builds RDF representation, which gives meaning to data of any domain and makes it easy for machines to process data and also semantically make use of the content (Fig. 2.5).

In the recent studies in the field of agriculture, there have been several websites or applications providing information about agricultural operations, probably each one operated by individual research organizations. Most of the applications provide automation in their respective fields. This field of applications always has some semantic information that gives a social aspect to society. Most of the farmers follow the traditional approaches to identification of symptoms, yield, plant growth, disease, etc. In this scenario, semantic applications (SAs) for agriculture will help

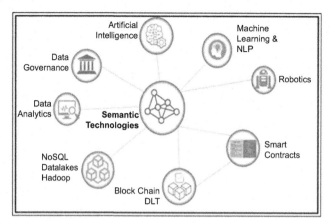

FIGURE 2.5

Semantic technology.

farmers in the automation of farming. The key challenge to building such SAs is the requirement of data that was semantically related to that particular field [9]. Moreover, the application also provides recommendations to farmers in a similar aspect. In a semantic vision, information can be readily interpreted by machines to do more tedious work for publishing, blogging, and many other areas.

2.2 Overview of semantic web protocol stack

A protocol stack is a set of standards that are used in a particular context. In SW protocol, a stack has layers and each layer has standards. Those are "semantics," "metadata," "ontologies," "proof," and "trust." The basic components of the SW are XML, RDF, RDF schema, and ontologies. XML is a framework used for writing semantic tags. The overall architecture of the SW protocol stack is presented in Fig. 2.6. It is divided into five layers, namely, syntactic, metadata, ontology, logic, and trust and proof layer [10]. Each layer performs a set of standards and technologies associated with it. The technical details of each layer with their representational capabilities are presented in the following subsections.

2.2.1 Syntactic layer

A syntactic layer represents the web documents as an abstraction of information that can contain a hierarchic structure of information. XML is the technology used to represent the hierarchic structure of information in this layer, to grant syntactic interoperability. XML uses a Uniform Resource Identifier and Internationalized Resource Identifiers for making everything language independent. These technologies, XML, URI, and IRI together, allow structuring, correct representation, and correct referencing of data for web applications.

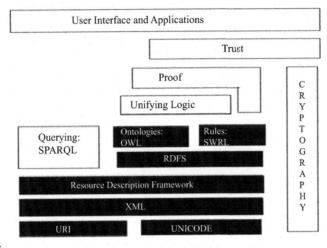

FIGURE 2.6

Semantic web layer architecture.

2.2.1.1 RDF/XML

XML was developed by the W3C to provide mechanisms to extract content and perform automated processing on documents. XML is a subset of the Standard Generalized Markup Language, a standard self-definition, i.e., structure of the data, which was embedded in the web pages. The main features of XML are platform independence, departure of the contents from their presentation, and the ability to determine structural rules of the document. A DTD (document type definition) structure is used to provide a well-defined meaning to XML elements and to validate the XML documents [11].

2.2.2 Metadata layer

In the next layer, metadata, the Semantic Web uses a simple XML-based data model, namely, RDF, for the description of resources and their types. In this model, data can be uniformly described with triples that are in specification with subject, predicate, and object. To make this machine understandable, it needs to define the meaning of the existing web content. The metadata layer is responsible for providing a well-defined meaning to the content of the web document. Metadata may be referred to as data about web pages, abstract resources, and everything that can be identified by any URI [11].

2.2.2.1 Resource description framework

In the SA development, resource description framework (RDF) plays a vital role in data representation and in an interchanging format. In this connection, RDF can act as a data model, in which it represents a resource that cannot be concerned with the context. The web relevance is also a kind of URI that can be given to the web resources headed with a specific URL. Besides, these particular URLs are the major building blocks of the statements to specified data. The data in the relational data-base management system can be represented in the form of records/tuples, whereas RDF formats are used for RDF data [12]. In this case, the RDF is comprised of tri-ples, where each triple consists of a subject, predicate, and object. These can be rep-resented by URIs, whereas objects can be URI or literal. Consequently, every literal can be used to model data values. Today, most of the RDF formats are available in the form of RDF/XML, RDF/JSON, and JSON-LD.

2.2.2.2 RDFs (RDF schema)

In the SW protocol stack, RDF is a vocabulary description language that provides metadata to the objects. A vocabulary is a term used repeatedly when defined, whereas metadata provides information about the URLs. Moreover, metadata along with vocabulary gives semantics for the data. Typically, these vocabulary terms can be stored in a namespace to make them easier to use, and they are stored using the RDF schema and Web Ontology Language (OWL) standards [11].

2.2.3 Ontology layer

Ontology adds a shareable and common semantics to the existing metadata without conveying any information about how to use concepts and relationships defined in the ontologies. Ontologies represent the most central component of the SW. It is a necessary framework for granting common metadata vocabulary understanding among applications. RDF schema is used to describe ontology formalisms. Ontology makes the resources machine processable and understandable, thereby fulfilling the main goal of the Semantic Web [13].

An ontology consists of individuals, properties, and classes. Concerning Protégé tool, resources are namely instances, slots, and classes. Classes are an object that hold the base description that follows hierarchic bases. Similarly, a class should contain a set of entities known as individuals or instances. These are represented in particular with class interests. Individuals can be referred to as instances of clas-ses. The relation between class and the individuals is represented as properties that have a binary relation. OWL classes are construed as assets that contain individuals. However, they are defined as a formal description that state specifically the require-ments for membership of the class [14].

2.2.4 Logic layer

The logic layer is added for requiring the meaning of the objects or classes and terms included in the ontology. OWL and Description Language (DL) are used for defining the formal semantics of the ontology. OWL is used for data representation, and DL is used for reasoning operators to reason about the data. SWRL is currently exploring the possibility of integrating rules and ontologies. Semantic Web Rule Language (SWRL) may be extended so a set of OWL axioms can be combined with a knowl-edge base.

2.2.4.1 SWRL rules (inference)

To suffice the needs of the Semantic Web, rule-based systems are widely used in various domains, such as medicine, compliance, engineering, diagnosis, law, and internet access and authentication. Rule Markup is a markup language utilized for sharing rules based on the WWW. Furthermore, it can build a hierarchy of rules and languages using RDF, XML, and OWL. In ontologies, SWRL rules for the inference mechanism are used for creating semantic knowledge rules. Concerning Protégé OWL development environment, SWRL rules work like mathematical first order and predicate logic.

2.2.5 Trust and proof layer

Trust and proof are the maximum goals for the development of the SW. All the technologies mentioned earlier are utilized for automatic integration of information, eliminating inconsistencies or incompleteness, and consolidating distributed information from the web. If they are conveyed by proper safety, authentication, and encryption mechanisms, then SW shall provide enough trust in the information required for the users.

2.3 **Challenge with semantic web**

Web services merely provide static information that allows some changes and effects on the WWW. SW enables users to trace, locate, choose, service, combine, and monitor web services automatically [15]. In this context, some of the key issues are as follows.

- **Web service discovery** is a semantic description of web services that can be recorded with intelligent agent proxies, public agents, cloud instances, and blockchain. These services may be migrated from one service to another service. In this context, various suitable webservices are defined for web service description languages. The XML-based registry standards are described as follows.

  (a) **Web Service Description Language** is determined in an XML format to specify abstract network services that provide data formats and protocol-based implementation. This cannot support semantic description services, as this is the main issue of web semantic language.

  (b) **Universal Description Discovery and Integration** is a web service repository consisting of XML schema definitions and their structures. It provides tModels (tagging messages) in the structures. The repository contains information to support services, business, and metadata categorical information. The key challenge is that it does not represent services capabilities.

- **Automatic web service invocation:** At present, web service requires human interaction during implementation. Consider an example: to buy a book in any online marketplace the website requires users to interact with their respective purchases. To complete the particular request for any service, multiple interactions are needed among users and web services. In this connection, automated web service abilities imply that a software broker can be extended on behalf of a user and infer that the user will experience the websites as well as the SNs [16]. These brokers interact with the web services that allow input data through computer interpretable application programming interfaces.

Agent-based distributed computing paradigm: SW uses ontologies to describe various web resources. These resources present semantics on the web that represent in a structured, logical, and semantic way. In a distributed paradigm, the network services are fully connected as an artificial neural network, and the services are provided to the users. In SW, distribution-based central agents and middle agents are required to help dynamic matchmaking among heterogeneous agents. These service providers can advertise their culpableness to the middle agents; then middle agents can subsequently store these advertisements. In the distributed agent-based computing environment, the following services are included:

- reduce the network capacity/load
- minimize network latency
- adapt dynamic migration in network speed
- maximize resource utilization

2.3.1 Semantics-based web search engines

- large-scale automatic search engines
- small-scale reviewer-based search engines
- **Web service composition and interoperation**: Web service composition is a mechanism for creating a new web service that makes use of existing services. Consequently, creating those service libraries can be automated in view of selecting and composing a novel web service to achieve the novel objective. The basics and significances of individual web services must be described formally. In the technology aspect, there is an automatic workflow generation [17] that can be used to generate automatic web services respectively.
- **Web service execution monitoring:** This is a performance monitoring for the web service. In the context of SW, ontology-based monitoring with inference rules is used to get more efficient results. The rules are inferred by logical reasoners that create the semantic knowledge to the services. In the long run of complex services, it is desirable to query each service execution status as well as its component in the service.

3. Knowledge representation and data processing management

3.1 Introduction to ontology

Ontologies can be widely used to represent the semantic knowledge of the particular domain of interest. They describe the information that is semantically contacted with resources. Moreover, the concepts in that particular domain make relationships. This provides web information that is more semantic and systematic. Recently, W3C provided a standard tool called Protégé used for developing a semantic knowledge base for SAs [18]. It facilitates a new rich set of operators, namely, union, intersection, and negation with the concepts. In the base semantic information, the SWRL rules create the inference mechanism to the resources in an efficient manner. Therefore, complex concepts can be built up in definitions in a simpler way. Hence, the reasoner can help to maintain the hierarchal data from classes to the data elements in a more significant way. This is practiced in real-world information for conversion to the Semantic Web. In a related study, the Protégé tool was the most recent development at Stanford University used to represent an information system using the OWL standards [13].

3.2 Ontology-based knowledge representation

Generally speaking, the study of semantics has been a challenging issue for years in the field of knowledge representation (KR), Natural Language Processor (NLP), and AI communities. Alongside, it encounters difficulties to access, present, and/or to manage the electronic information on the web. Even so, this requires data collection and representation that enables software products (agents) to provide intelligent access to various heterogeneous and distributed machines. However, the Semantic Web movement brings the issue to the web, whereas the RDF framework is designed for describing and interchanging metadata [19]. Although, several key challenges in knowledge representation over ontologies are assisting applications to maintain the interoperability of metadata and to enhance the better precision in resource discovery compared to full-text search.

Of late, KR and reasoning can be a promising field of NLP and AI, whereas KR can be determined to represent the information of social applications. Subsequently, ontology-based knowledge representation (OBKR) and reasoning mechanisms can be importantly responsible for semantic knowledge environments. However, knowledge-based reasoning approaches are derived through new conclusions for the preparation of dynamic nondeterministic solutions. The KR can be derived based on ontology and the methods for reasoning that contribute to paths planning. Furthermore, a detailed comparison of several planning domain modeling languages, planners, ontology editors, and robot simulation tools are depicted as follows.

Fig. 2.7 depicts a sample OBKR that describes individual instances and functions of the domain representation utilizing unary and binary predicates. Moreover, this representation allows knowledge capturing, processing, sharing, reuse, and communication. Nevertheless, the OMLs are coined through adopting the base of ontology-based knowledge management including RDFS, OIL, DAML + OIL, OWL, RDF, OWL − DL, and OWL − Lite. Eventually, the low-level description logic services bring support especially for reasoning and remedies for inconsistency, whereas ontology is based on DL, a low-level KR method. Thereafter, several representations of ontology KR including semantic knowledge, spatial knowledge, and temporal knowledge representations are discussed as follows [20] (Fig. 2.8)

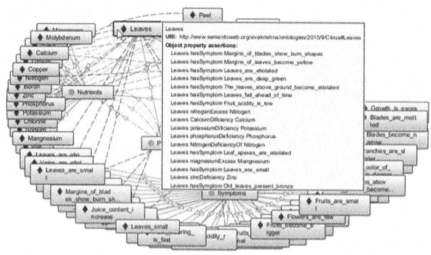

FIGURE 2.7

Sample ontology for citrus fruits.

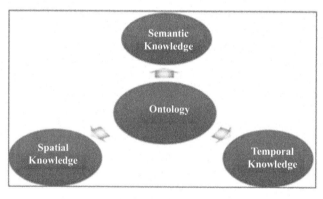

FIGURE 2.8

Ontology-based knowledge representations.

3.2.1 Semantic knowledge representations

These representations consist of the generic knowledge of concepts and relations among resources.

Herein, the term *"semantics"* refers to *"signs"* and *"things"* refers to *"entities."* Semantic knowledge is responsible for the instruction and detailed information especially needed for the performance of the system. In this connection, robotic systems' new inferences will enable robots to perform enormous tasks and robots will have adequate knowledge to identify not only the deterministic path but also perform better in the artificial world. Moreover, knowledge utilization of robotics in the learning process is more efficient, whereas task planning has a sequence set of ordered actions to achieve a high-level goal. In recent years, most of the AI-based applications have been using this type of knowledge representation. Further, it can also be used in many machine learning, blockchain applications [20].

3.2.2 Spatial knowledge representation

Spatial knowledge represents the conception of space and shape of the objects. The object information is stored in mage. Each image consists of pixels that are stored in a matrix or vector representation. This data element processing will take more complex task management. In artificial machines, image processing is highly challenging in nature. Most applications in machine learning modules like image captioning, face detection, and object detection depend on special knowledge. In this context, the challenge is motion detection and object identification, and actions on a particular object always depend on the image spatial representation and feature extraction.

3.2.3 Temporal knowledge representation

Temporal knowledge represents information containing qualitative and quantitative elements. Quantitative data measure the values, expressed as a number, whereas qualitative data are numeric variables. In a related study of ontology temporal language, OWL is extended to have the temporal semantics. This information will order in SW reason and represent the temporal information. In robotic systems, this type of information requires a sophisticated knowledge in the environment for a processing task.

3.3 Query processing in semantic web databases

Today's Semantic Web datasets are becoming progressively larger, containing over one billion triples. It is well known for the web of linked data to allow people to build vocabularies, to create data stores on the web, and write rules for handling data. In a Semantic Web application, the key challenge is to access RDF and OWL data sources. SPARQL provides data processing and retrieves data from RDF and supports querying of Machine-Readable Data File (MRDF) graphs and OWL data.

3.3.1 Query representation and processing

Significant queries in SPARQL are query representation, processing, and evaluation [21]. In a general query, representation is a statement processed by the query execution engine. Later, this statement is evaluated and syntactically parsed. In the following section, we presented some of the query models, query processing, query parsing, and rewriting and execution planning.

A. Query models: In the past, relational databases were used for storing information and SQL queries were used in the retrieval process. In Ref. [22], the authors describe a novel model named Query Graph Model, and later, Hartig and Reese [23] modified it with SPARQL queries for an appropriate evaluation scheme with its operations. These models are interpreted as a directed graph where nodes denote operators and edges represent data flow. In a related study, SPARQL Query Graph Pattern Model was focused on query optimization in SW. This model generates query graph patterns instead of operation themselves as in SPARQL Query Graph Model (SQGM).

B. Query processing: It is performed by the SPARQL execution engine, which was evaluated by the frontend compiler. Moreover, these processes can be validated and consequently compilation and query processing can also be done accordingly. Besides, various components including an optimizer, evaluation engine, and translation have been used. In execution, a pipeline can be initialized in connection to the plan from the front end.

C. Query parsing and rewriting: Based on the W3C recommendation, these standard methods are widely used to execute syntactic and lexical analysis accordingly. SQGPM processes input streams and then transform into short forms in queries. Later in the second step is query rewriting, which expects some of the queries to be represented optimally. These contain deceits, constant expressions, ineffective conditions, redundancies, etc. Therefore, the goal is to achieve better performance in normalizing queries and also requires checking the applicability of every operation according to SPARQL semantics before use of preserving query equivalency [24].

D. Execution plan generation: It is a visual representation of the operations that are performed by the execution engine to return results from the query. An evaluation plan is viewed by the query optimizer and query engine. It decides in which order data is retrieved, type of joins, data size, filter, aggregation, and key constraint issues. Moreover, the dynamic programming approach is used to search for all possible paths in search space. This applies group patterns in SQGPM models and also provides a better overall plan [25] (Fig. 2.9).

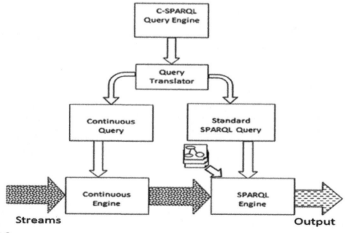

FIGURE 2.9

SPARQL query processing.

3.3.2 Evaluation of SPARQL queries

When evaluating the SPARQL queries the compiler finishes the compilation and then passes to the query processing plan. This plan can be essential to transform into a performance processing engine for its evaluation. The flowing components are used while executing the query in the query execution engine.

A. Data representation: Data representation in SPARQL is in terms of RDF format. It is typically redundant because many triples share the same object in the context that is practiced. To reduce the redundancy for storing the RDF data for every instance, there needs a unique string in the memory. This can be represented in the RDF database.

B. Transformation of the query execution plan: To transform the execution plan query, it is deserialized, and operators must be switched by connection. However, the approach implies parallel exploitation typically in the form of hierarchic bases.

C. Implementation of query plan operators: In the implementation process of the query, some of the operations are required, namely, index scan, filter, sort, merge join, nested loops join, hash, join, optional joins, distinct, and explicit parallelization of nested loops join.

D. Explicit parallelization with nested loop joins: The process of joining operation parallelization of query results creates optimal/efficient evaluation. This reduces the query execution time by the execution engine.

3.4 **Key challenges in query processing**

Some of the key challenges in query processing are as follows.

A. Parallel databases: The web-related data is stored in various servers or data centers, so query with the parallel databases should be optimal. In this context, various optimization approaches have been developed to efficiently process in SW databases. In a multicore computer, a massive performance boost occurs by parallel computing. In related studies, the parallel algorithms for join computations of SPARQL query significantly speed up querying large sets of data linked in the semantic world [26].

B. Intertransaction parallelism: Different transactions running in parallel have been a standard practice for decades. Multilevel transaction management provides variants of nested transactions. These are hierarchic base models where the nodes are executions of operators on an abstract level, and the edges imply a sequence of operations. In a related study, transaction storage is an important aspect of parallelism in a shared disk environment [27]. In this context, multilevel transaction-based approaches are used in the present world of database systems in intertransactions.

C. Intratransaction parallelism: Queries of a transaction may be executed in parallel, provided they do not interfere among themselves and they do not interact with the external world. Since these conditions are met rather rarely, this kind of parallelism is seldom exploited except for experiments.

D. Interoperator parallelism: It is a form of parallelism in the evaluation of database queries that were used in a concurrent process in multicore processing to improve the throughput of the system. In multicore systems, interleaving forms one transaction to others. This creates a system with inconsistent and deadlock issues. To overcome independent behavior, arrangement to turn in parallel makes it relatively easier [28].

E. Inference: Data contains given facts, which are expressed explicitly. Inference rules are mostly regeneration of new rules learning by themselves. For example, SWRL rules are the inference rules for the RDF databases where inference is based on the rule decision systems. These systems create an artificial learning process to the machines as well as providing better results in the query processing.

F. Logical optimization: An algebra for SPARQL queries is to define the semantics of query evaluation. In the relational model, logical optimization gives efficient results by making use of operations like selection, projection, join, nested queries, subqueries, etc. In this context of SW, logical query optimization deals with query optimizers, which estimate possible query plans. Finally, in cost estimation, histograms for estimating the cardinality of operators result as a basis [29].

G. Physical optimization: Different algorithms figure out the logical operators like UNION, FIND, OPT, SORT, AND, etc. This operator cost depends on which the result estimates. From the logical point of view, the best-estimated execution times gives a better operator. In the resulting query, it is necessary that which operator has been used for evaluation is also desired for the implementation.

H. Streams: Data streams are becoming an important concept and are used in many applications. Processing of data streams needs a steaming engine that starts query processing. This ability is especially important for real-time computation for distribution and parallelism of long relay transmission of data streams. In a related study of stream processing, eBay auctions, which are based on the RDF stream engine and can flexibly analyze eBay auctions, were studied. Using a monitoring system, users can easily monitor the eBay auction information of interest, analyze the behavior of buyers and sellers, predict the tendency of auctions, and make more favorable decisions [30].

I. Visual query languages: The social web has gradually become widespread and essential in the creation of collective intelligence. In visual representation, RDF query languages play a positive role in the social investigation for data extraction. Moreover, the related study of visual query systems will help social data in the semantic formulation of the query [31]. Furthermore, this system also supports many other functionalities, namely, precise suggestions and refining existing queries.

4. Social information networks applications and architecture

4.1 Introduction to social network architecture and applications

Facebook: It is a social network of friends, family, and other business houses and organizations, which are connected with a network. Moreover, these users are sharing their personal information, posting content videos, photos, and event listings in the backbone network.

Facebook network architecture is a centralized network, which consists of components, namely, BGP protocol, sFlow controller, and Open/R. BGP is a gateway routing protocol that allows traffic from one control to the other. sFlow controller samples the network devices to feed inactive traffic control based on the current demand. Open/R provides routing in both interior routing and messaging in the network. Moreover, agents are running on network devices to provide inspector general of police and messaging functionality. In central control management, Local Service Provider (LSP) agents play a critical interfacing role between each data center pair. Fig. 2.10 shows the Facebook network architecture in which primary elements are interconnected.

Facebook software design uses a crossbreed model for traffic engineering. This model consists of both distributed control agents and a central controller. It allows traffic control in an intelligent computing path and handles network failures. The central distributed system signals between Open/R agents on the network nodes. In EBB, there are local agents to redirect the traffic from Open/R agents to other

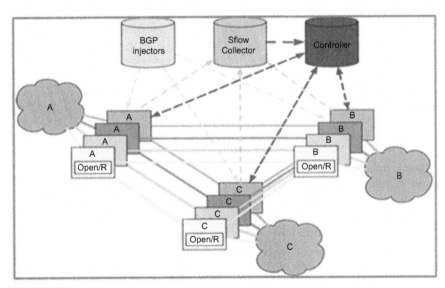

FIGURE 2.10

Facebook network architecture (control stack).

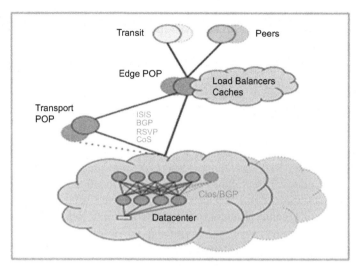

FIGURE 2.11

Twitter network architecture.

agents. Moreover, network planes are used to split physical topology into parallel topology to improve network performance.

Twitter: It is an American microblogging and social network service on which users connect to each other by sharing their secret messages known as "tweets." The application always gives the best possible experience for users in technology and hardware efficiency.

In late 2010, Twitter network architecture was designed, and at that time, there were some issues related to scalability and service in hosting with COLO technology. Later, the application was running with the POP network with hundreds of thousands of services. Fast forward a few years and technologies were converted to Cols with Border Gateway Protocol (BGP). A BGP network that had a live network handled complexity with minimal impact on services (Fig. 2.11) [32].

4.2 Challenges in new approaches

■ **Data center traffic:** Highly scalable architectural data centers are built in such a way that they can handle forklift migration and network traffic. Moreover, Twitter provides a centralized distributed environment that is used for the user application in a better growing network in real time [33]. Nevertheless, data center traffic deals with some issues:

- traffic trends toward the upper end of the designed capacity
- impacts in hosted and cloud environment like design decisions and network operations
- no temporary change or workaround

■ **Backbone traffic:** It is mainly focused on dynamic growth in the network traffic between servers in the data centers, but there is a challenge in the backbone network. In this regard, to solve the issue, the network administrators used multiprotocol label switching with a resource reservation protocol that allowed automatic bandwidth based on the network demands [34]. The key challenges with backbone traffic are as follows.
  • efficient maintenance in large LSPs in reducing processing capacity
  • traffic in low-priority backend traffic
  • maintain the available bandwidth

■ **Storage:** There are millions of tweets sent each day by the users, which are stored, cached, managed, assisted, and analyzed. This needs a consequent infrastructure for messaging and storage in a storage area network [35]. The messaging and storage services are as follows.
  • **Hadoop:** Hadoop clusters run for both Compute and HDFS (Hadoop Distributes File System). It has multiple clusters storing over 500 PB of data. The clusters were divided into groups, namely, processing, data warehouse, real time, and cold storage. Each cluster has its responsibility to segregate the workload
  • **Manhattan:** It is a backend for tweets, direct messages, and Twitter accounts. It runs quite a few clusters in different use cases like multitenant, traffic pattern in reads/writes, etc. These handle millions of queries in a second and give efficient results in the data center.
  • **Blob store:** It is an API that allows an application to serve data objects called blobs. These are much larger than the size allowed for objects in the data storage service. Twitter user information like images, videos, and large files are stored in the form of objects.
  • **Cache:** It is temporary storage, in which the recent query results were stored. This type of storage device mainly is used in fast processing in data storage centers. A few technologies like Redis and Twemcache are used in the Twitter network architecture.

Most of this architecture is mainly focused on cluster-based cache techniques.

4.3 Challenges in the network architecture
• incremental deployment with upgraded features
• dynamic growth in network capacity
• efficient network routing leans by multiprotocol label switching segment

4.4 Challenges in software design
• improved visibility of traffic demand
• minimize latency-insensitive traffic
• bandwidth allocation on the explicit contract between the network and services

- minimize path congestion for latency-insensitive traffic
- inconsistent debugging issues in the control of bad LSPs

4.5 Challenges of social network application development

Today, building a social application becomes a challenging factor to facilitate a better experience for the audience. To develop such social applications, some of the key elements are performance, security, design, and personalization [36].

- **Performance:** It can be one of the most impact factors of application development. To provide better performance, an application must be developed with the vision of future growth in a scalable environment. When people experience the application, it should be high performance in terms of processing, responding, and navigating. The key performance issues are memory leaks, response time, throughput, queuing delay, waiting for the delay, sessional delay, etc.
- **Security:** To maintain the control of privacy, almost every user must interact with a secure environment. Preserving users' security and integrity helps to achieve end user business goals and scale the revenues. The key challenges in security are vulnerable storage, insecure data points, cache vulnerable, feeble cryptography, etc. For example, Facebook has its own privacy settings to contact only friend's lists and posts to control the privacy issues in applications.
- **Design:** Most of the users are experienced with UI design (user interface). Sometimes, due to effortlessness, it could create stickiness. When designing an application the core features should be clear and the interface understandable.
- **Personalization:** Many users have various perceptions according to the content and bids. Apply learning filters to retain the public on the same platform. Learning filters allow remembering the individual's activities and suggest content that is relevant to the feature.

5. Conclusion

Nowadays, social networks play a vital role in the WWW. In the present social network, technology changes from year to year. There are many applications rapidly growing data and social connections. The key challenge is the social connection in the applications is not semantically connected by the network. In this chapter, we have addressed social networks with semantic technology and also its applications along with architectural and design issues. In this context the new approach is with Semantic Web and the important aspect of knowledge representations, data models with ontologies, query processing with SPARQL, and proof and trust issues. Moreover, we have given a brief introduction to various key components, namely, RDF, ontology tools, execution process, and application development issues.

References

[1] B.-L. Tim, The World Wide Web: A Very Short Personal History, 1998. http://www.w3.org/People/Berners-Lee/ShortHistory.html.

[2] K.B. Maged, N, W. Steve, The emerging Web 2.0 social software: an enabling suite of sociable technologies in health and health care education, Health Inf. Libr. J. (2007) 2–23.

[3] B.-L. Tim, H. James, L. Ora, The Semantic Web A new form of Web content that is meaningful to computers will unleash a revolution of new possibilities, Sci. Am. Fea. Artic. Semantic Web (May 2001).

[4] B.D. Deebak, F. Al-Turjman, A novel community-based trust aware recommender systems for big data cloud service networks, Sustainable Cities Soc. 2020, 102274.

[5] G. Appel, L. Grewal, R. Hadi, et al., The future of social media in marketing, J. Acad. Market. Sci. 48 (2020) 79–95, https://doi.org/10.1007/s11747-019-00695-1.

[6] J. Sun, X. Ren, C.J. Anumba, Analysis of knowledge-transfer mechanisms in construction project cooperation networks, J. Manage. Eng. 35 (2) (2019) 04018061.

[7] R. Priambodo, R. Satria, User behavior pattern of mobile online social network service, in: International Conference on Cloud Computing and Social Networking (ICCCSN), Bandung, West Java, 2012, 2012, pp. 1–4, https://doi.org/10.1109/ICCCSN.2012.6215732.

[8] A. Sofroniou, Technological Integrations, Lulu. com, 2019.

[9] R.Q. Grafton, D. Garrick, A. Manero, T.N. Do, The water governance reform framework: Overview and applications to Australia, Mexico, Tanzania, USA, and Vietnam, Water 11 (1) (2019) 137.

[10] D. Sell, L. Cabral, E. Motta, J. Domingue, FarshadHakimpour, R. Pacheco, A semantic web-based architecture for analytical tools, in: Proceedings of the Seventh IEEE 2005 International Conference on E-Commerce Technology (CEC'05), July 2005, pp. 347–354, 19-22.

[11] L. Terán, Dynamic Profiles for Voting Advice Applications: An Implementation for the 2017 Ecuador National Elections, Springer, 2019.

[12] Y. Ding, C. Wang, WangMaoguang, The Research on data semantic description framework using RDF/XML, in: Proceedings of the First International IEEE Conference on Semantics, Knowledge, and Grid (SKG 2005), Beijing, Nov. 2005, p. 138, 27-29.

[13] S.K. Dasari, K.R. Chintada, M. Patruni, Flue-cured tobacco leaves classification: a generalized approach using deep convolutional neural networks, in: Cognitive Science and Artificial Intelligence, Springer, Singapore, 2018, pp. 13–21.

[14] Natalya F. Noy, P.D., Monica Crubezy, Ph.D., Ray W F., Ph.D., Holger K., Ph.D., Samson W. Tu, MS, J. Vendetti, BS Mark A. M., MD, Ph.D.: "Protege-2000: an open-source ontology-development and knowledge-acquisition environment" AMIA 2003 Symposium Proceedings – Page 953.

[15] Y. Ding, D. Fensel, Ontology library systems: the key for successful ontology reuse, in: The First Semantic Web Working Symposium, Springer-Verlag Heidelberg, Heidelberg, 2001, pp. 93–112.

[16] X. Fan, P. Zhang, J. Zhao, Transformation of relational database schema to semantics web model, in: Second International Conference on Communication Systems, Networks and Applications, Hong Kong vol. 1, 2010, pp. 379–384. June 29 2010-July 1 2010.

[17] S. Lu, M. Dong, F. Fotouhi, The Semantic Web: opportunities and challenges for next-generation Web applications, Inf. Res. 7 (4) (2002).

[18] A. Akbari, J. Rosell, et al., Ontological physics-based motion planning for manipulation, in: Emerging Technologies & Factory Automation, ETFA, 2015 IEEE 20th Conference on, IEEE, 2015, pp. 1–7.

[19] L. Belouaer, M. Bouzid, A.-I. Mouaddib, Ontology-based spatial planning for human-robot interaction Temporal Representation and Reasoning, in: TIME, 2010 17th International Symposium on, IEEE, 2010, pp. 103–110.

[20] L. Dong, D.N. Seed, H. Li, W.R. Flynn IV, C. Wang, G. Lu, C.M. Mladin, U.S. Patent No. 10,432,449, U.S. Patent and Trademark Office, Washington, DC, 2019.

[21] D. Bednarek, J. Dokulil, J. Yaghob, F. Zavoral, Using methods of parallel semi-structured data processing for semantic web, in: 3rd International Conference on Advances in Semantic Processing, SEMAPRO, IEEE Computer Society Press, 2009, pp. 44–49.

[22] H. Pirahesh, J.M. Hellerstein, W. Hasan, Extensible/rule-based query rewrite optimization in a starburst, SIGMOD Rec. 21 (June 1992) 39–48.

[23] O. Hartig, R. Heese, The SPARQL query graph model for query optimization, in: E. Franconi, M. Kifer, W. May (Eds.), The Semantic Web: Research and Applications, Ser. Lecture Notes in Computer Science vol. 4519, Springer Berlin/Heidelberg, 2007, pp. 564–578.

[24] M. Cermak, J. Dokulil, F. Zavoral, SPARQL compiler for bobox, in: Fourth International Conference on Advances in Semantic Processing, 2010, pp. 100–105.

[25] H.O. Eggen, K.A. Hole, An Evaluation of Join-Strategies in a Distributed MySQL Plugin Architecture (Master's Thesis, NTNU), 2019.

[26] C. Qin, H. Eichelberger, K. Schmid, Enactment of adaptation in data stream processing with latency implications—A systematic literature review, Inf. Software Technol. 111 (2019) 1–21.

[27] F. Morvan, A. Hameurlain, Dynamic memory allocation strategies for parallel query execution, in: SAC '02: Proceedingsof the 2002 ACM Symposium on Applied Computing, ACM, New York, NY, USA, 2002, pp. 897–901.

[28] A.N. Wilschut, J. Flokstra, P.M.G. Apers, Parallel evaluation of multi-joint queries, in: SIGMOD'95: Proceedings of the 1995 ACM SIGMOD International Conference on Management of Data, ACM, New York, NY, USA, 1995, pp. 115–126.

[29] M.N. Mami, D. Graux, H. Thakkar, S. Scerri, S. Auer, J. Lehmann, The Query Translation Landscape: A Survey, 2019 arXiv preprint arXiv:1910.03118.

[30] S. Katragadda, R. Gottumukkala, S. Venna, N. Lipari, S. Gaikwad, M. Pusala, M. Bayoumi, VAStream: a visual analytics system for fast data streams, in: Proceedings of the Practice and Experience in Advanced Research Computing on Rise of the Machines (Learning), 2019, pp. 1–8.

[31] J. Groppe, S. Groppe, A. Schleifer, Visual Query System for Analyzing Social Semantic Web, 2011, pp. 217–220, https://doi.org/10.1145/1963192.1963293.

[32] V.K. Mishra, Software Defined Networks, Momentum Press, 2019.

[33] B.D. Deebak, A Secure-Ware System for Web Server: Ensuring Platform Interoperability, Security, Privacy, Usability and Functionality, National Acad. Sci. Lett. 40 (3) (2017) 157–160.

[34] A. Hbaieb, M. Khemakhem, M.B. Jemaa, A survey and taxonomy on virtual data center embedding, J. Supercomput. 75 (10) (2019) 6324–6360.

[35] A.T. Fadi, D.B. David, Seamless authentication: For IoT-big data technologies in smart industrial application systems, IEEE Transactions on Industrial Informatics, IEEE (2020), https://doi.org/10.1109/TII.2020.2990741.

[36] F. Paternò, C. Santoro, End-user development for personalizing applications, things, and robots, Int. J. Hum. Comput. Stud. 131 (2019) 120−130.

Cyber security in mobile social networks

Fadi Al-Turjman[1], Ramiz Salama[2]

[1]*Research Center for AI and IoT, Near East University, Nicosia, Mersin, Turkey;* [2]*Computer Engineering Department, Research Center for AI and IoT, Near East University, Nicosia, Mersin, Turkey*

1. Introduction

Cyber breaches have dominated headlines as attacks targeting more and more users have grown dramatically. Mobile technology seems to be an easy target for cyber criminals who seek financial gain by stealing credit card data or personal information that can be resold or used for extortion. Criminal networks have reaped immense profits and are able to invest into investigating and developing more sophisticated methods and skills, which are then available through online forums for anyone to purchase. Meanwhile, there is no dependable and effective defense mechanism for mobile technology to fend off such attacks. The lack of knowledge or training to face today's security challenges remains an issue with the users of mobile devices.

Mobile security, or more specifically mobile device security, has become increasingly important in mobile computing. Of particular concern is the security of personal and business information now stored on smartphones.

More and more users and businesses use smartphones to communicate, but also to plan and organize their users' work and also private life. Within companies, these technologies are causing profound changes in the organization of information systems, and therefore they have become the source of new risks. Indeed, smartphones collect and compile an increasing amount of sensitive information to which access must be controlled to protect the privacy of the user and the intellectual property of the company.

All smartphones, as computers, are preferred targets of attacks. These attacks exploit weaknesses inherent in smartphones that can come from the communication mode, like Short Message Service (SMS, aka text messaging), Multimedia Messaging Service (MMS), Wi-Fi, Bluetooth, and GSM, the de facto global standard for mobile communications. There are also exploits that target software vulnerabilities in the browser or operating system. And some malicious software relies on the weak knowledge of an average user.

Security countermeasures are being developed and applied to smartphones, from security in different layers of software to the dissemination of information

Security in IoT Social Networks. https://doi.org/10.1016/B978-0-12-821599-9.00003-0
Copyright © 2021 Elsevier Inc. All rights reserved.

to end users. There are good practices to be observed at all levels, from design to use, through the development of operating systems, software layers, and downloadable apps.

2. Challenges in mobile app security

2.1 Device fragmentation

Mobile application testing needs to cover a multiplicity of mobile devices with different capabilities, features, and limitations. Identification of security vulnerabilities specific to devices makes performance testing a difficult task. The testing team cannot test releases as fast as the development team is producing them, so they are becoming a bottleneck in the release process. This also leads to the production of low-quality apps. Most of the apps are made in iOS, Android, or Windows environments. But there are different versions of each operating system (OS) that have a different set of vulnerabilities. Testing of the app on each version is time-consuming and requires the application tester to be aware of the loopholes.

2.2 Tools for mobile automation testing

A reasonable approach to fragmentation requires the use of automation testing. But traditional testing tools like Selenium or QuickTest Professional (QTP) were not designed with cross-platform in mind. So automation tools for mobile apps and web applications are different. While many test automation and testing tools for mobile have emerged, there is a dearth of full-fledged standard tools that can cater to every step of the security testing. The common mobile automation testing tools are Appium, Robotium, and Ranorex.

2.3 Weak encryptions

A mobile app can accept data from all kinds of sources. In the absence of sufficient encryption, attackers could modify inputs such as cookies and environment variables. Attackers can bypass the security when decisions on authentication and authorization are made based on the values of these inputs. Recently, hackers targeted Starbucks mobile users to siphon money out of their Starbucks mobile app. Starbucks confirmed that its app was storing usernames, email addresses, and passwords in clear text. This allowed anyone with access to the phone to see passwords and usernames just by connecting the phone to a PC.

2.4 Weak hosting controls

When creating their first mobile applications, businesses often expose server-side systems that were previously inaccessible to outside networks. The servers on which your app is hosted should have security measures to prevent unauthorized users from accessing data. This includes your own servers and the servers of any third-party systems your app may be accessing. It is important for the back-end services to be secured against malicious attacks. Thus, all APIs should be verified, and proper security methods should be employed ensuring access to authorized personnel only.

2.5 Insecure data storage

In most of the popular apps, consumers simply enter their passwords once when activating the payment portion of the app and use it again and again to make unlimited purchases without having to reinput their password or username. In such cases, user data should be secure and usernames, email addresses, and passwords should be encrypted. For example, in 2012 a flaw in Skype data security allowed hackers to open the Skype app and dial arbitrary phone numbers using a simple link in the contents of an email. Design apps in such a way that critical information such as -contact details, passwords, and credit card numbers do not reside directly on a device. If they do, they must be stored securely.

3. What is a mobile threat?

Like viruses and spyware that can infect your PC, there are a variety of security threats that can affect mobile devices. We divide these mobile threats into several categories: application-based threats, web-based threats, network-based threats, and physical threats.

3.1 Application-based threats

Downloadable applications can present many types of security issues for mobile devices. "Malicious apps" may look fine on a download site, but they are specifically designed to commit fraud. Even some legitimate software can be exploited for fraudulent purposes. Application-based threats generally fit into one or more of the following categories.

Malware is software that performs malicious actions while installed on your phone. Without your knowledge, malware can make charges to your phone bill, send unsolicited messages to your contact list, or give an attacker control over your device.

Spyware is designed to collect or use private data without your knowledge or approval. Data commonly targeted by spyware includes phone call history, text messages, user location, browser history, contact list, email, and private photos. This stolen information could be used for identity theft or financial fraud.

Privacy threats may be caused by applications that are not necessarily malicious, but gather or use sensitive information (e.g., location, contact lists, personally identifiable information) that is necessary to perform their function.

Vulnerable applications are apps that contain flaws that can be exploited for malicious purposes. Such vulnerabilities allow an attacker to access sensitive information, perform undesirable actions, stop a service from functioning correctly, or download apps to your device without your knowledge.

3.2 Web-based threats

Because mobile devices are constantly connected to the internet and frequently used to access web-based services, web-based threats pose persistent issues for mobile devices.

Phishing scams use email, text messages, Facebook, and Twitter to send you links to websites that are designed to trick you into providing information like passwords or account numbers. Often these messages and sites are very different to distinguish from those of your bank or other legitimate sources.

Drive-by downloads can automatically download an application when you visit a web page. In some cases, you must take action to open the downloaded application, while in other cases the application can start automatically.

Browser exploits take advantage of vulnerabilities in your mobile web browser or software launched by the browser such as a Flash player, PDF reader, or image viewer. Simply by visiting an unsafe web page, you can trigger a browser exploit that can install malware or perform other actions on your device.

3.3 Network threats

Mobile devices typically support cellular networks as well as local wireless networks (Wi-Fi, Bluetooth). Both of these types of networks can host different classes of threats.

Network exploits take advantage of flaws in the mobile operating system or other software that operates on local or cellular networks. Once connected, they can install malware on your phone without your knowledge.

Wi-Fi sniffing intercepts data as it is traveling through the air between the device and the Wi-Fi access point. Many applications and web pages do not use proper security measures, sending unencrypted data across the network that can be easily read by someone who is grabbing data as it travels.

3.4 Physical threats

Mobile devices are small, valuable, and we carry them everywhere with us, so their physical security is also an important consideration.

Lost or stolen devices are one of the most prevalent mobile threats. The mobile device is valuable not only because the hardware itself can be resold on the black market, but more importantly because of the sensitive personal and organization information it may contain.

4. Types of wireless attacks

Wireless attacks can come at you through different methods. For the most part, you need to worry about Wi-Fi. Some methods rely on tricking users, others use brute force, and some look for people who do not bother to secure their network. Many of these attacks are intertwined with each other in real-world use. Here are some of the kinds of attacks you could encounter.

4.1 Packet sniffing

When information is sent back and forth over a network, it is sent in what we call packets. Since wireless traffic is sent over the air, it is very easy to capture. Quite a lot of traffic (FTP, HTTP, SNMP, etc.) is sent in the clear, meaning that there is no encryption, so files are in plain text for anyone to read. So using a tool like Wireshark allows you to read data transfers in plain text! This can lead to stolen passwords or leaks of sensitive information quite easily. Encrypted data can be captured as well, but it is obviously much harder for an attacker to decipher the encrypted data packets.

4.2 Rogue access point

When an unauthorized access point (AP) appears on a network, it is refereed to as a rogue access point. These can range from an employee who does not know better to a person with ill intent. These APs represent a vulnerability to the network because they leave it open to a variety of attacks. These include vulnerability scans for attack preparation, ARP poisoning, packet captures, and distributed denial of service (DDoS) attacks.

4.3 Password theft

When communicating over wireless networks, think of how often you log into a website. You send passwords out over the network, and if the site does not use SSL or TLS, that password is sitting in plain text for an attacker to read. There are even ways to get around those encryption methods to steal the password. I will talk about this with man-in-the-middle attacks.

4.4 Man-in-the-middle attack

It is possible for hackers to trick communicating devices into sending their transmissions to the attacker's system. Here they can record the traffic to view later (like in packet sniffing) and even change the contents of files. Various types of malware can be inserted into these packets, email content could be changed, or the traffic could be dropped so that communication is blocked.

4.5 Jamming

There are a number of ways to jam a wireless network. One method is flooding an AP with deauthentication frames. This effectively overwhelms the network and prevents legitimate transmissions from getting through. This attack is a little unusual because there probably is not anything in it for the hacker. One of the few examples of how this could benefit someone is through a business jamming their competitor's Wi-Fi signal. This is highly illegal (as are all these attacks), so businesses would tend to shy away from it. If they got caught, they would be facing serious charges.

4.6 War driving

War driving comes from an old term called war dialing, where people would dial random phone numbers in search of modems. War driving is basically people driving around looking for vulnerable APs to attack. People will even use drones to try and hack APs on higher floors of a building. A company that owns multiple floors around 10 stories up might assume nobody is even in range to hack their wireless, but there is no end to the creativity of hackers!

4.7 Bluetooth attacks

There are a variety of Bluetooth exploits out there. These range from annoying pop up messages to full control over the victim's Bluetooth-enabled device.

4.8 WEP/WPA attacks

Attacks on wireless routers can be a huge problem. Older encryption standards are extremely vulnerable, and it is pretty easy to gain the access code in this case. Once someone is on your network, you have lost a significant layer of security. APs and routers are hiding your IP address from the broader internet using Network Address Translation (unless you use IPv6, but that is a topic for another day). This effectively hides your private IP address from those outside your subnet and helps prevent outsiders from being able to directly attack you. The keyword there is that it *helps* prevent the attacks; it does not stop them completely.

Another thing of which to take note is that our mobile devices are at risk whenever they connect to public Wi-Fi. Whether you use a phone, tablet, or laptop, accessing an insecure network is putting a target on your data. Understand the risks or consider using a virtual private network (VPN).

5. Attacks based on communication

A smartphone user is exposed to various threats when they use their phone. These threats can disrupt the operation of the smartphone and transmit or modify user data. So applications must guarantee privacy and integrity of the information they handle. In addition, since some apps could themselves be malware, their functionality and activities should be limited (for example, restricting the apps from accessing location information via GPS, blocking access to the user's address book, preventing the transmission of data on the network, sending SMS messages that are billed to the user, etc.).

There are three prime targets for attackers:

- Data: smartphones are devices for data management, and they may contain sensitive data like credit card numbers, authentication information, private information, and activity logs (calendar, call logs);
- Identity: smartphones are highly customizable, so the device or its contents can easily be associated with a specific person. For example, every mobile device can transmit information related to the owner of the mobile phone contract, and an attacker may want to steal the identity of the owner of a smartphone to commit other offenses;
- Availability: attacking a smartphone can limit access to it and deprive the owner of its use.

There are a number of threats to mobile devices, including annoyance, stealing money, invading privacy, propagation, and malicious tools.

- Botnets: attackers infect multiple machines with malware that victims generally acquire via email attachments or from compromised applications or websites. The malware then gives hackers remote control of "zombie" devices, which can then be instructed to perform harmful acts;
- Malicious applications: hackers upload malicious programs or games to third-party smartphone application marketplaces. The programs steal personal information and open backdoor communication channels to install additional applications and cause other problems;
- Malicious links on social networks: an effective way to spread malware where hackers can place Trojans, spyware, and backdoors;
- Spyware: hackers use this to hijack phones, allowing them to hear calls, see text messages and emails, as well as track someone's location through GPS updates.

The sources of these attacks are the same actors found in the nonmobile computing space.

- Professionals, whether commercial or military, who focus on the three targets mentioned before. They steal sensitive data from the general public, as well as undertake industrial espionage. They will also use the identity of those attacked to achieve other attacks;
- Thieves who want to gain income through data or identities they have stolen. The thieves will attack many people to increase their potential income;
- Black hat hackers who specifically attack availability. Their goal is to develop viruses and cause damage to the device. In some cases, hackers have an interest in stealing data on devices;
- Gray hat hackers who reveal vulnerabilities. Their goal is to expose vulnerabilities of the device. Gray hat hackers do not intend to damage the device or steal data.

5.1 **Consequences**

When a smartphone is infected by an attacker, the attacker can attempt several things:

- The attacker can manipulate the smartphone as a zombie machine, that is to say, a machine with which the attacker can communicate and send commands that will be used to send unsolicited messages (spam) via SMS or email [2];
- The attacker can easily force the smartphone to make phone calls. For example, one can use the API (library that contains the basic functions not present in the smartphone) PhoneMakeCall by Microsoft, which collects telephone numbers from any source such as yellow pages, and then call them [2]. But the attacker can also use this method to call paid services, resulting in a charge to the owner of the smartphone. It is also very dangerous because the smartphone could call emergency services and thus disrupt those services [2];
- A compromised smartphone can record conversations between the user and others and send them to a third party [2]. This can cause user privacy and industrial security problems;
- An attacker can also steal a user's identity, usurp their identity (with a copy of the user's SIM card or even the telephone itself), and thus impersonate the owner. This raises security concerns in countries where smartphones can be used to place orders, view bank accounts, or are used as an identity card
- The attacker can reduce the utility of the smartphone, by discharging the battery [3]. For example, they can launch an application that will run continuously on the smartphone processor, requiring a lot of energy and draining the battery. One factor that distinguishes mobile computing from traditional desktop PCs is their limited performance. Frank Stajano and Ross Anderson first described this form of attack, calling it an attack of "battery exhaustion" or "sleep deprivation torture";
- The attacker can prevent the operation and/or starting of the smartphone by making it unusable [4]. This attack can either delete the boot scripts, resulting in a phone without a functioning OS, or modify certain files to make it unusable (e.g., a script that launches at startup that forces the smartphone to restart), or even embed a startup application that would empty the battery;
- The attacker can remove the personal (photos, music, videos, etc.) or professional data (contacts, calendars, notes) of the user.

5.2 **Attack based on SMS and MMS**

Some attacks derive from flaws in the management of SMS and MMS.

Some mobile phone models have problems in managing binary SMS messages. It is possible, by sending an ill-formed block, to cause the phone to restart, leading to the DDoS attacks. If a user with a Siemens S55 received a text message containing a Chinese character, it would lead to a DDoS [5]. In another case, while the standard requires that the maximum size of a Nokia Mail address is 32 characters, some Nokia phones did not verify this standard, so if a user enters an email address over 32 characters, that leads to complete dysfunction of the email handler and puts it out of commission. This attack is called "curse of silence" A study on the safety of the SMS infrastructure revealed that SMS messages sent from the internet can be used to perform a DDoS attack against the mobile telecommunications infrastructure of a big city. The attack exploits the delays in the delivery of messages to overload the network.

Another potential attack could begin with a phone that sends an MMS to other phones, with an attachment. This attachment is infected with a virus. Upon receipt of the MMS, the user can choose to open the attachment. If it is opened, the phone is infected, and the virus sends an MMS with an infected attachment to all the contacts in the address book. There is a real-world example of this attack: the virus Commwarrior uses the address book and sends MMS messages including an infected file to recipients. A user installs the software, as received via MMS message. Then, the virus began to send messages to recipients taken from the address book.

6. Attacks based on vulnerabilities in software applications and hardware

6.1 Other attacks are based on flaws in the operating system or applications on the phone

6.1.1 Web browser

The mobile web browser is an emerging attack vector for mobile devices. Just as common web browsers, mobile web browsers are extended from pure web navigation with widgets and plug-ins or are completely native mobile browsers.

Jailbreaking the iPhone with firmware 1.1.1 was based entirely on vulnerabilities on the web browser [6]. As a result, the exploitation of the vulnerability described here underlines the importance of the web browser as an attack vector for mobile devices. In this case, there was a vulnerability based on a stack-based buffer overflow in a library used by the web browser (Libtiff).

A vulnerability in the web browser for Android was discovered in October 2008. As with the aforementioned iPhone vulnerability, it was due to an obsolete and vulnerable library. A significant difference with the iPhone vulnerability was Android's sandboxing architecture, which limited the effects of this vulnerability to the web browser process.

Smartphones are also victims of classic piracy related to the web: phishing, malicious websites, software that run in the background, etc. The big difference is that smartphones do not yet have strong antivirus software available.

6.1.2 Operating system

Sometimes it is possible to overcome the security safeguards by modifying the operating system itself. As real-world examples, this section covers the manipulation of firmware and malicious signature certificates. These attacks are difficult.

In 2004, vulnerabilities in virtual machines running on certain devices were revealed. It was possible to bypass the bytecode verifier and access the native underlying operating system. The results of this research were not published in detail. The firmware security of Nokia's Symbian Platform Security Architecture (PSA) is based on a central configuration file called SWIPolicy. In 2008, it was possible to manipulate the Nokia firmware before it is installed, and in fact in some downloadable versions of it, this file was human readable, so it was possible to modify and change the image of the firmware [7]. This vulnerability has been solved by an update from Nokia.

In theory, smartphones have an advantage over hard drives since the OS files are in ROM, so they cannot be changed by malware. However, in some systems, it was possible to circumvent this: in the Symbian OS it was possible to overwrite a file with a file of the same name [7]. On the Windows OS, it was possible to change a pointer from a general configuration file to an editable file.

When an application is installed, the signing of this application is verified by a series of certificates. One can create a valid signature without using a valid certificate and add it to the list [8]. In the Symbian OS, all certificates are in the directory: c:\resource\swicertstore\dat. With firmware changes explained earlier, it is very easy to insert a seemingly valid but malicious certificate.

6.2 Attacks based on hardware vulnerabilities

6.2.1 Electromagnetic waveforms

In 2015, researchers at the French government agency Agence nationale de la sécurité des systèmes d'information (ANSSI) demonstrated the capability to trigger the voice interface of certain smartphones remotely by using "specific electromagnetic waveforms" [1]. The exploit took advantage of antenna properties of headphone wires while plugged into the audio output jacks of the vulnerable smartphones and effectively spoofed audio input to inject commands via the audio interface.

6.2.2 Juice jacking

Juice jacking is a physical or hardware vulnerability specific to mobile platforms. Utilizing the dual purpose of the USB charge port, many devices have been susceptible to having data exfiltrated from, or malware installed onto, a mobile device by utilizing malicious charging kiosks set up in public places or hidden in normal charge adapters.

6.2.3 Jailbreaking and rooting

Jailbreaking is also a physical access vulnerability, in which mobile device users initiate to hack into the devices to unlock it, and exploit weaknesses in the operating system. Mobile device users take control of their own device by jailbreaking it, and customize the interface by installing applications, changing system settings that are not allowed on the devices. Thus, this allows them to tweak the mobile devices operating systems processes and run programs in the background, so devices are being exposed to a variety of malicious attacks that can lead to compromise of important private data.

7. Portability of malware across platforms

There is a multitude of malware. This is partly due to the variety of operating systems on smartphones. However, attackers can also choose to make their malware target multiple platforms, and malware can be found that attacks an OS but is able to spread to different systems.

To begin with, malware can use runtime environments like Java virtual machine or the .NET Framework. They can also use other libraries present in many operating systems. Other malwares carry several executable files to run in multiple environments, and they utilize these during the propagation process. In practice, this type of malware requires a connection between the two operating systems to use as an attack vector. Memory cards can be used for this purpose, or synchronization software can be used to propagate the virus.

7.1 Resource monitoring in the smartphone

When an application passes the various security barriers, it can take the actions for which it was designed. When such actions are triggered, the activity of a malicious application can be sometimes detected if one monitors the various resources used on the phone. Depending on the goals of the malware, the consequences of infection are not always the same; all malicious applications are not intended to harm the devices on which they are deployed. The following sections describe different ways to detect suspicious activity.

7.2 Battery

Some malware is aimed at exhausting the energy resources of the phone. Monitoring the energy consumption of the phone can be a way to detect certain malware applications.

7.3 Memory usage

Memory usage is inherent in any application. However, if one finds that a substantial proportion of memory is used by an application, it may be flagged as suspicious.

7.4 **Network traffic**

On a smartphone, many applications are bound to connect via the network, as part of their normal operation. However, an application using a lot of bandwidth can be strongly suspected of attempting to communicate a lot of information and disseminate data to many other devices. This observation only allows a suspicion, because some legitimate applications can be very resource-intensive in terms of network communications, the best example being streaming video.

7.5 **Services**

One can monitor the activity of various services of a smartphone. During certain moments, some services should not be active, and if one is detected, the application should be suspected. For example, the sending of an SMS when the user is filming video: this communication does not make sense and is suspicious; malware may attempt to send SMS while its activity is masked.

The various points mentioned before are only indications and do not provide certainty about the legitimacy of the activity of an application. However, these criteria can help target suspicious applications, especially if several criteria are combined.

7.6 **Network surveillance**

Network traffic exchanged by phones can be monitored. One can place safeguards in network routing points to detect abnormal behavior. As the mobile device's use of network protocols is much more constrained than that of a computer, expected network data streams can be predicted (e.g., the protocol for sending an SMS), which permits detection of anomalies in mobile networks.

7.7 **Spam filters**

As is the case with email exchanges, we can detect a spam campaign through means of mobile communications (SMS, MMS). It is therefore possible to detect and minimize this kind of attempt by filters deployed on network infrastructure that is relaying these messages.

7.8 **Encryption of stored or transmitted information**

Because it is always possible that data exchanged can be intercepted, communications, or even information storage, can rely on encryption to prevent a malicious entity from using any data obtained during communications. However, this poses the problem of key exchange for encryption algorithms, which requires a secure channel.

7.9 **Telecom network monitoring**

The networks for SMS and MMS exhibit predictable behavior, and there is not as much liberty compared with what one can do with protocols such as TCP or UDP. This implies that one cannot predict the use made of the common protocols of the web; one might generate very little traffic by consulting simple pages, rarely, or generate heavy traffic by using video streaming. On the other hand, messages exchanged via mobile phone have a framework and a specific model, and the user does not, in a normal case, have the freedom to intervene in the details of these communications. Therefore, if an abnormality is found in the flux of network data in the mobile networks, the potential threat can be quickly detected.

7.10 **Manufacturer surveillance**

In the production and distribution chain for mobile devices, it is the responsibility of manufacturers to ensure that devices are delivered in a basic configuration without vulnerabilities. Most users are not experts and many of them are not aware of the existence of security vulnerabilities, so the device configuration as provided by manufacturers will be retained by many users. Next are listed several points that manufacturers should consider.

7.11 **Remove debug mode**

Phones are sometimes set in a debug mode during manufacturing, but this mode must be disabled before the phone is sold. This mode allows access to different features, not intended for routine use by a user. Due to the speed of development and production, distractions occur, and some devices are sold in debug mode. This kind of deployment exposes mobile devices to exploits that utilize this oversight.

7.12 **Default settings**

When a smartphone is sold, its default settings must be correct, and not leave security gaps. The default configuration is not always changed, so a good initial setup is essential for users. There are, for example, default configurations that are vulnerable to DDoS attacks.

7.13 **Security audit of apps**

Along with smart phones, appstores have emerged. A user finds themselves facing a huge range of applications. This is especially true for providers who manage appstores because they are tasked with examining the apps provided, from different points of view (e.g., security, content). The security audit should be particularly cautious, because if a fault is not detected, the application can spread very quickly within a few days and infect a significant number of devices.

7.14 Detect suspicious applications demanding rights

When installing applications, it is good to warn the user against sets of permissions that, grouped together, seem potentially dangerous, or at least suspicious. Frameworks like such as Kirin, on Android, attempt to detect and prohibit certain sets of permissions.

7.15 Revocation procedures

Along with appstores appeared a new feature for mobile apps: remote revocation. First developed by Android, this procedure can remotely and globally uninstall an application, on any device that has it. This means the spread of a malicious application that managed to evade security checks can be immediately stopped when the threat is discovered.

7.16 Avoid heavily customized systems

Manufacturers are tempted to overlay custom layers on existing operating systems, with the dual purpose of offering customized options and disabling or charging for certain features. This has the dual effect of risking the introduction of new bugs in the system, coupled with an incentive for users to modify the systems to circumvent the manufacturer's restrictions. These systems are rarely as stable and reliable as the original and may suffer from phishing attempts or other exploits.

7.17 User awareness

Much malicious behavior is allowed by the carelessness of the user. Smartphone users were found to ignore security messages during application installation, especially during application selection, checking application reputation, reviews, and security and agreement messages. From simply not leaving the device without a password, to precise control of permissions granted to applications added to the smartphone, the user has a large responsibility in the cycle of security: to not be the vector of intrusion. This precaution is especially important if the user is an employee of a company that stores business data on the device. Detailed subsequently are some precautions that a user can take to manage security on a smartphone.

A recent survey by internet security experts BullGuard showed a lack of insight into the rising number of malicious threats affecting mobile phones, with 53% of users claiming that they are unaware of security software for smartphones. A further 21% argued that such protection was unnecessary, and 42% admitted it had not crossed their mind ("Using APA," 2011). These statistics show consumers are not concerned about security risks because they believe it is not a serious problem. The key here is to always remember smartphones are effectively handheld computers and are just as vulnerable.

7.18 **Being skeptical**

A user should not believe everything that may be presented, as some information may be phishing or attempting to distribute a malicious application. It is therefore advisable to check the reputation of the application that they want to buy before actually installing it.

7.19 **Permissions given to applications**

The mass distribution of applications is accompanied by the establishment of different permissions mechanisms for each operating system. It is necessary to clarify these permissions mechanisms to users, as they differ from one system to another, and they are not always easy to understand. In addition, it is rarely possible to modify a set of permissions requested by an application if the number of permissions is too great. But this last point is a source of risk because a user can grant rights to an application far beyond the rights it needs. For example, a note taking application does not require access to the geolocation service. The user must ensure the privileges are required by an application during installation and should not accept the installation if requested rights are inconsistent.

8. **How to secure your mobile device**

The rapidly changing technology and portability of mobile devices have forced people to rely heavily on those products. With their increased functionalities, mobile devices carry out a number of our day-to-day activities, such as surfing the web, booking appointments, setting up reminders, sharing files, instant messaging, video calling, and even mobile banking.

Given all these functionalities, mobile devices are vulnerable to online threats and are also susceptible to physical attacks due to their portability. Some of the security threats include malware specifically designed for mobile devices, e.g., worms and spyware, unauthorized access, phishing, and theft.

But not all is lost. Here are some practical steps that will help you minimize the exposure of your mobile device to digital threats.

8.1 **Use strong passwords/biometrics**

Strong passwords coupled with biometric features, such as fingerprint authenticators, make unauthorized access nearly impossible. Your passwords should be eight or more characters long and contain alphanumeric characters. If your mobile device allows two-factor authentication, do not hesitate to use it. You do not want to be subject to unforeseen attacks.

The complexities of your passwords in other apps might tempt you to store them in a similar way a browser does, that is, using the "remember me feature."

This feature should be avoided at all costs since it only increases the chances of your password getting spoofed. Alternatively, if you lose your device, another person might gain full access to it.

Furthermore, do not forget to change your password from time to time (at least every 3 months).

8.2 Ensure public or free Wi-Fi is protected

Everybody loves free Wi-Fi, especially when their data plan is inexpensive. But cheap can turn expensive in a very devastating manner. That is because most of the free Wi-Fi points are not encrypted. These open networks allow malicious people to eavesdrop on the network traffic and easily get your passwords, usernames, and other sensitive information. That threat is not going anywhere anytime soon, either. According to eVoice Australia, a premier provider of virtual telecommunications solutions, Wi-Fi hacking will constitute a growing risk for many in 2017.

To protect against Wi-Fi hacking, use applications that secure your connection or at least tell you the status of the Wi-Fi to which you are connected. WPA (Wi-Fi Protected Access) is more secure than WEP (Wired Equivalent Privacy).

As a matter of being cautious, you should also turn off wireless connectivity (Wi-Fi and Bluetooth) when you are not using them. Not only will this help avoid automatic connection to unencrypted networks, but it also saves your battery.

8.3 Utilize VPN

If you are not sure about the security status of the network to which you are connected, using a VPN client is mandatory. A VPN will enable you to connect to a network securely. At the same time, any browsing activity you do on the public Wi-Fi will be shielded from prying eyes.

It is also useful when accessing sites that are less secure. Non-HTTPS sites are visible to anyone who knows how to use networking and vulnerability tools. These sites are prone to MITM (Man-in-the-middle) attacks, which pave a way to eavesdropping and password sniffing. You really need to have a new mindset when it comes to fighting cybercrime.

8.4 Encrypt your device

Most mobile devices are bundled with a built-in encryption feature. Encryption is the process of making data unreadable. Decryption is converting the unreadable data into normal data. This is important in case of theft, and it prevents unauthorized access. You simply need to locate this feature on your mobile device and enter a password to encrypt your device.

This process may take time depending on the size of your data. The bigger the data, the more time it will take. Most importantly, you need to remember the encryption password because it is required every time you want to use your mobile device.

Also, as a fail-safe, consider backing up your data since some mobile devices will automatically erase everything if the wrong encryption password is entered incorrectly after a number of times.

8.5 Install an antivirus application

The files you download and the apps you install on your mobile device might be packed with malicious code. Once launched, this code could send your data to hackers, thereby making you insecure and robbing you of your privacy. To avoid that, installing a reputable antivirus application will guarantee your security.

Some antivirus applications also offer more functionalities, such as erasing your data if you lose your mobile device, tracking and blocking unknown callers who might be a threat, and telling you which applications are not safe.

In addition, they offer to clear your browsing history and delete cookies. Cookies are small software tokens that store your login information that might be leaked if someone malicious gets.

9. Mobile security threats you should take seriously

Mobile security is at the top of every company's worry list these days, and for good reason. Nearly all workers now routinely access corporate data from smartphones, and that means keeping sensitive info out of the wrong hands is an increasingly intricate puzzle. The stakes, suffice it to say, are higher than ever: The average cost of a corporate data breach is a whopping $3.86 million, according to a 2018 report by the Ponemon Institute. That is 6.4% more than the estimated cost just 1 year earlier.

While it is easy to focus on the sensational subject of malware, the truth is that mobile malware infections are incredibly uncommon in the real world, with your odds of being infected significantly less than your odds of being struck by lightning, according to one estimate. Malware currently ranks as the least common initial action in data breach incidents, in fact, coming in behind even physical attacks in Verizon's 2019 Data Breach Investigations Report. That is thanks to both the nature of mobile malware and the inherent protections built into modern mobile operating systems.

The more realistic mobile security hazards lie in some easily overlooked areas, all of which are only expected to become more pressing as we make our way through 2019.

9.1 Data leakage

It may sound like a diagnosis from the robot urologist, but data leakage is widely seen as being one of the most worrisome threats to enterprise security in 2019. Remember those almost nonexistent odds of being infected with malware? Well, when it comes to a data breach, companies have a nearly 28% chance of experiencing at least one incident in the next 2 years, based on Ponemon's latest research, odds of more than one in four, in other words.

What makes the issue especially vexing is that it often is not nefarious by nature; rather, it is a matter of users inadvertently making ill-advised decisions about which apps are able to see and transfer their information.

"The main challenge is how to implement an app vetting process that does not overwhelm the administrator and does not frustrate the users," says Dionisio Zumerle, research director for mobile security at Gartner. He suggests turning to mobile threat defense (MTD) solutions, products like Symantec's Endpoint Protection Mobile, CheckPoint's SandBlast Mobile, and Zimperium's zIPS Protection. Such utilities scan apps for "leaky behavior," Zumerle says, and can automate the blocking of problematic processes.

Of course, even that will not always cover leakage that happens as a result of overt user error, something as simple as transferring company files onto a public cloud storage service, pasting confidential info in the wrong place, or forwarding an email to an unintended recipient. That is a challenge the healthcare industry is currently struggling to overcome: According to specialist insurance provider Beazley, "accidental disclosure" was the top cause of data breaches reported by healthcare organizations in the third quarter of 2018. That category combined with insider leaks accounted for nearly half of all reported breaches during that time span.

For that type of leakage, data loss prevention tools may be the most effective form of protection. Such software is designed explicitly to prevent the exposure of sensitive information, including in accidental scenarios.

9.2 Social engineering

The tried-and-true tactic of trickery is just as troubling on the mobile front as it is on desktops. Despite the ease with which one would think social engineering cons could be avoided, they remain astonishingly effective.

A staggering 91% of cybercrime starts with email, according to a 2018 report by security firm FireEye. The firm refers to such incidents as "malware-less attacks," since they rely on tactics like impersonation to trick people into clicking dangerous links or providing sensitive info. Phishing, specifically, grew by 65% over the course of 2017, the company says, and mobile users are at the greatest risk of falling for it because of the way many mobile email clients display only a sender's name, making it especially easy to spoof messages and trick a person into thinking an email is from someone they know or trust.

Users are actually three times more likely to respond to a phishing attack on a mobile device than a desktop, according to an IBM study, in part because a phone is where people are most likely to first see a message. Verizon's latest research supports that conclusion and adds that the smaller screen sizes and corresponding limited display of detailed information on smartphones (particularly in notifications, which frequently now include one-tap options for opening links or responding to messages) can also increase the likelihood of phishing success.

Beyond that, the prominent placement of action-oriented buttons in mobile email clients and the unfocused, multitasking-oriented manner in which workers tend to use smartphones amplify the effect, and the fact that the majority of web traffic is generally now happening on mobile devices only further encourages attackers to target that front.

It is not just email anymore, either: As enterprise security firm Wandera noted in its latest mobile threat report, 83% of phishing attacks over the past year took place outside the inbox, in text messages or in apps like Facebook Messenger and WhatsApp along with a variety of games and social media services.

What is more, while only a single-digit percentage of users actually click on phishing-related links—anywhere from 1% to 5%, depending on the industry, according to Verizon's most current data—earlier Verizon research indicates those gullible guys and gals tend to be repeat offenders. The company notes that the more times someone has clicked on a phishing campaign link, the more likely they are to do it again in the future. Verizon has previously reported that 15% of users who are successfully phished will be phished at least one more time *within the same year.*

"We do see a general rise in mobile susceptibility driven by increases in mobile computing overall [and] the continued growth of BYOD work environments," says John "Lex" Robinson, information security and antiphishing strategist at PhishMe, a firm that uses real-world simulations to train workers on recognizing and responding to phishing attempts.

Robinson notes that the line between work and personal computing is also continuing to blur. More and more workers are viewing multiple inboxes—connected to a combination of work and personal accounts—together on a smartphone, he notes, and almost everyone conducts some sort of personal business online during the workday. Consequently, the notion of receiving what appears to be a personal email alongside work-related messages does not seem at all unusual on the surface, even if it may in fact be a ruse.

The stakes only keep climbing higher. Cybercrooks are apparently now even using phishing to try to trick folks into giving up two-factor authentication codes designed to protect accounts from unauthorized access. Turning to hardware-based authentication—either via dedicated physical security keys like Google's Titan or Yubico's YubiKeys or via Google's on-device security key option for Android phones—is widely regarded as the most effective way to increase security and decrease the odds of a phishing-based takeover.

According to a study conducted by Google, New York University, and UC San Diego, even just on-device authentication can prevent 99% of bulk phishing attacks and 90% of targeted attacks, compared to a 96% and 76% effectiveness rate for those same types of attacks with the more phishing-susceptible 2FA codes.

9.3 Wi-Fi interference

A mobile device is only as secure as the network through which it transmits data. In an era where we are all constantly connecting to public Wi-Fi networks, that means our info often is not as secure as we might assume.

Just how significant of a concern is this? According to research by Wandera, corporate mobile devices use Wi-Fi almost three times as much as they use cellular data. Nearly a quarter of devices have connected to open and potentially insecure Wi-Fi networks, and 4% of devices have encountered a man-in-the-middle attack—in which someone maliciously intercepts communication between two parties—within the most recent month. McAfee, meanwhile, says network spoofing has increased "dramatically" as of late, and yet less than half of people bother to secure their connection while traveling and relying on public networks.

"These days, it is not difficult to encrypt traffic," says Kevin Du, a computer science professor at Syracuse University who specializes in smartphone security. "If you don't have a VPN, you're leaving a lot of doors on your perimeters open."

Selecting the right enterprise-class VPN, however, is not so easy. As with most security-related considerations, a tradeoff is almost always required. "The delivery of VPNs needs to be smarter with mobile devices, as minimizing the consumption of resources—mainly battery—is paramount," Gartner's Zumerle points out. An effective VPN should know to activate only when absolutely necessary, he says, and not when a user is accessing something like a news site or working within an app that is known to be secure.

9.4 Out-of-date devices

Smartphones, tablets, and smaller connected devices—commonly known as the internet of things (IoT)—pose a new risk to enterprise security in that unlike traditional work devices, they generally do not come with guarantees of timely and ongoing software updates. This is true particularly on the Android front, where the vast majority of manufacturers are embarrassingly ineffective at keeping their products up to date—both with OS updates and with the smaller monthly security patches between them—as well as with IoT devices, many of which are not even designed to get updates in the first place.

"Many of them do not even have a patching mechanism built in, and that is becoming more and more of a threat these days," Du says.

Increased likelihood of attack aside, an extensive use of mobile platforms elevates the overall *cost* of a data breach, according to Ponemon, and an abundance of work-connected IoT products only causes that figure to climb further. The IoT is "an open door," according to cybersecurity firm Raytheon, which sponsored research showing that 82% of IT professionals predicted that unsecured IoT devices would cause a data breach—likely "catastrophic"—within their organization.

Again, a strong policy goes a long way. There *are* Android devices that do receive timely and reliable ongoing updates. Until the IoT landscape becomes less of a wild west, it falls upon a company to create its own security net around them.

9.5 **Cryptojacking attacks**

A relatively new addition to the list of relevant mobile threats, cryptojacking is a type of attack where someone uses a device to mine for cryptocurrency without the owner's knowledge. If all that sounds like a lot of technical mumbo-jumbo, just know this: The cryptomining process uses your company's devices for someone else's gain. It leans heavily on *your* technology to do it, which means affected phones will probably experience poor battery life and could even suffer from damage due to overheating components.

While cryptojacking originated on the desktop, it saw a surge on mobile from late 2017 through the early part of 2018. Unwanted cryptocurrency mining made up a third of all attacks in the first half of 2018, according to a Skybox Security analysis, with a 70% increase in prominence during that time compared to the previous half-year period. And mobile-specific cryptojacking attacks absolutely exploded between October and November of 2017, when the number of mobile devices affected saw a 287% surge, according to a Wandera report.

Since then, things have cooled off somewhat, especially in the mobile domain, a move aided largely by the banning of cryptocurrency mining apps from both Apple's iOS App Store and the Android-associated Google Play Store in June and July, respectively. Still, security firms note that attacks continue to see some level of success via mobile websites (or even just rogue ads on mobile websites) and through apps downloaded from unofficial third-party markets.

Analysts have also noted the possibility of cryptojacking via internet-connected set-top boxes, which some businesses may use for streaming and video casting. According to security firm Rapid7, hackers have found a way to take advantage of an apparent loophole that makes the Android Debug Bridge—a command-line tool intended only for developer use—accessible and ripe for abuse on such products.

For now, there's no great answer—aside from selecting devices carefully and sticking with a policy that requires users to download apps only from a platform's official storefront, where the potential for cryptojacking code is markedly reduced—and realistically, there's no indication that most companies are under any significant or immediate threat, particularly given the preventative measures being taken across the industry. Still, given the fluctuating activity and rising interest in this area over the past months, it is something well worth being aware of and keeping an eye on in the future.

9.6 **Poor password hygiene**

You would think we would be past this point by now, but somehow, users still are not securing their accounts properly, and when they are carrying phones that contain both company accounts *and* personal sign-ins, that can be particularly problematic.

A recent survey by Google and Harris Poll found just over half of Americans, based on the survey's sample, reuse passwords across multiple accounts. Equally concerning, nearly a third are not using 2FA (or do not *know* if they are using it, which might be a little worse). Only a quarter of people are actively using a password manager, which suggests the vast majority of folks probably do not have particularly strong passwords in most places, since they are presumably generating and remembering them on their own.

Things only get worse from there: According to a 2018 LastPass analysis, a full half of professionals use the same passwords for both work and personal accounts. And if *that* is not enough, an average employee shares about six passwords with a coworker over the course of his or her employment, the analysis found.

Lest you think this is all much ado about nothing, in 2017, Verizon found that weak or stolen passwords were to blame for more than *80%* of hacking-related breaches in businesses. From a mobile device in particular—where workers want to sign in quickly to various apps, sites, and services—think about the risk to your organization's data if even just one person is sloppily typing in the same password they use for a company account into a prompt on a random retail site, chat app, or message forum. Now combine *that* risk with the aforementioned risk of Wi-Fi interference, multiple it by the total number of employees in your workplace, and think about the layers of likely exposure points that are rapidly adding up.

Perhaps most vexing of all, most people seem completely oblivious to their oversights in this area. In the Google and Harris Poll survey, 69% of respondents gave themselves an "A" or "B" at effectively protecting their online accounts, despite subsequent answers that indicated otherwise. Clearly, you cannot trust a user's own assessment of the matter.

9.7 Physical device breaches

Last but not least is something that seems especially silly but remains a disturbingly realistic threat: A lost or unattended device can be a major security risk, especially if it does not have a strong PIN or password and full data encryption.

Consider the following: In a 2016 Ponemon study, 35% of professionals indicated their work devices had no mandated measures in place to secure accessible corporate data. Worse yet, nearly half of those surveyed said they had no password, PIN, or biometric security guarding their devices, and about two-thirds said they did not use encryption. Sixty-eight percent of respondents indicated they sometimes shared passwords across personal and work accounts accessed via their mobile devices.

Things do not seem to be getting any better. In its 2019 mobile threat landscape analysis, Wandera found that 43% of companies had at least one smartphone in their roster without any lock screen security. And among users who did set up passwords or PINs on their devices, the firm reports, many opted to use the bare-minimum four-character code when given the opportunity.

The take-home message is simple: Leaving the responsibility in users' hands is not enough. Do not make assumptions; make policies. You will thank yourself later.

10. Tips for securing Wi-Fi

Now that you do not trust anything on the internet anymore, let's build that confidence back up. There are a lot of ways to make yourself less susceptible to wireless attacks.

10.1 Use WPA2 security

This takes enough work to crack that most hackers will look for an easier target. Make sure WPS is turned off!

10.2 Minimize your network reach

Try to position your router in the center of your home or building. There are tools available to measure the reach of your network, and you can adjust the signal level. Try to make it so that the signal beyond your walls is degraded enough that it is not useable. You may also consider using a directional antenna if central placement is not an option.

10.3 Use firewalls

Make sure your APs firewall is enabled. If you can afford a hardware firewall and feel you need the extra security, go ahead and install one. Household networks generally can get away with the standard router firewall and operating system firewalls.

10.4 Use a VPN on open networks

If you really must use public Wi-Fi, set up a VPN. Most smartphones have this capability. You can set one up on your PC. This allows you to communicate through an encrypted tunnel back to your home or office. You can even send web traffic through a VPN.

10.5 Update software and firmware

Keep your system up to date with the latest patches, and make sure any online applications you use are updated as well. Check for AP firmware updates related to security flaws, and implement them as soon as possible. Remember to follow best practices for network modification to ensure you do not interrupt a critical task. Check out your updates in a test lab to make sure that they do not interfere with an important application. Do not perform updates during normal operating hours if possible, and if you must update during work hours make sure everyone is aware that network connectivity could slow down or be cut off temporarily while you work.

10.6 Use strong passwords

I recommend you use at least a 15-character password. Use a mix of upper/lowercase letters, numbers, and symbols. Again, do not make it easy. Is the only capital letter at the start? Is there an exclamation at the end? Are there any words in there? These are common bad password practices, and hackers love them.

10.7 Change the login credentials

Make sure you change the administrative login credentials. This is often something like admin/admin or admin/password by default.

10.8 Disable your SSID (service set identifier) broadcast

This is not a security measure. The right tools will still find your network's SSID (this is the name of your network in case you did not know). However, there's a small chance it could help your network fly under the radar.

10.9 Enable MAC filtering

Again, MAC filtering is not security. A knowledgeable hacker knows how to monitor your network and copy the MAC address of a connected device. They can then spoof their own MAC to appear as an authorized device to gain access. However, this is another annoyance for them to deal with.

It is a good idea to monitor your network connections to look for unusual activity. If you have an Android phone, you can use a free network IP scanner to see the IP addresses of connected devices. Desktops can use something like the nmap tool. For a home network with few devices, you want to find out what your devices IP addresses currently are and see if there are any that do not match. Be aware that if your Wi-Fi uses DHCP (automatically assigned IP's) that these could change over time.

There are a lot of ways for hackers to come after your data, but taking these simple precautionary measures and proactively monitoring for threats can make a world of difference.

11. Challenges and open issues

Mobile networks have left the "walled garden." A privileged, closed, and isolated ecosystem, which is under the full control of mobile carriers, used proprietary protocols and has minimal security risks due to restricted user access. With the introduction of 5G, Long Term Evolution (LTE) networks, and the IEEE standardization of mobile networks, the secure, "walled garden" days are over. Mobile networks are becoming very similar to common IP-based networks. However, while organizations have years of experience and knowledge in defending against cyber threats in common IP networks, they are years behind in terms of accumulated knowledge in defending mobile networks. In the following, we highlight major issues in this area:

- Migration has not been met with equal funding; providers have not invested in security at the same rate as capacity.
- Mobile service providers must adapt to upcoming changes in the threat landscape and be prepared to ensure network availability.
- Since new threats may have potential for catastrophe, the mobile network providers should adapt security programs and procedures to withstand those threats, while assuring the same SLA and preventing any major service outages.

11.1 Key technical recommendations

As with any other network security program, there is no magic solution that will eliminate all threats. The most effective and robust solution should be the product of a careful risk assessment process, which analyzes all business critical "weak spots" and implements the proper technical and procedural solutions to compound them.

The various technical solutions involved in such a solution include the following:

- Deep packet inspection (DPI): deep packet inspection solutions that were designed to be implemented on mobile networks, on network gateways, in the access networks, and in the core network itself;
- Behavioral-based mitigation: since many threats are 0-day attacks or use legitimate transactions to misuse resources, the only way they can be detected and stopped is by behavioral analysis techniques. These techniques will allow normal network traffic to pass through while mitigating abnormal traffic patterns;
- VOIP protections: solutions that will protect VOIP signaling infrastructure from being exploited from both internal and external sources;
- DNS protections: specific DNS-oriented protections that will be able to detect and effectively block any massive abuse of the various DNS services found in the mobile networks;
- Signaling proxies: that will handle all signaling traffic from a single point and make sure the network will not be overwhelmed from the signaling traffic rates.

12. Conclusion

Like any data communication network, mobile networks contain a range of security threats. Though some threats are easy to identify and mitigate, others are illusive, due to the unique structure and complexity of mobile networks. The transformation to fully IP-based mobile networks involves a transition period, during which existing security vulnerabilities will be exposed to a substantially larger audience. Attackers can easily generate attacks targeting mobile endpoints, overlapping network services between mobile and other networks (such as DNS), and even the core network itself. A full-blown attack on a mobile network has the potential for catastrophic results that affect multiple audiences. In today's hyperconnected world, it could be viewed as a national infrastructure attack.

This study examines the major information security threats relating to mobile services and solutions to these threats from the service developer's perspective. Research methods employed include interviews with enterprises, literature searches, expert opinions, and extensive rounds of commentary. The fact that information security threats also concern mobile services and should be given serious consideration is the most important finding of the study. However, this does not mean information security issues would pose an obstacle to the development or introduction of mobile services. All information security issues need to be addressed at the very outset of the service development process. Methods and technologic solutions that may also be utilized in mobile services have already been developed. Sets of instructions safeguarding the security of actions and processes are less readily available.

The major information security threats facing developers of mobile services include the complexity of technologic solutions, the illegal copying of content and programs, threats posed by the internet, the different levels of various players in the service development process, and threats involving the identification of service users and servers and the confidentiality of information.

References

[1] U.D. Ulusar, F. Al-Turjman, G. Celik, An overview of Internet of things and wireless communications, in: In2017 International Conference on Computer Science and Engineering (UBMK), IEEE, October 2017, pp. 506–509.

[2] P. Ahonen, J. Eronen, J. Holappa, J. Kajava, T. Kaksonen, K. Karjalainen, R. Savola, Information security threats and solutions in the mobile world, VTT Res. Notes (2005) 1–108.

[3] X. Wang, P. Yi, Security framework for wireless communications in smart distribution grid, IEEE Trans. Smart Grid 2 (4) (2011) 809–818.

[4] Rasheed, U., Soofi, A. A., Sarwar, M. U., & Khan, M. I. A Study on the Security of Mobile Devices, Network and Communication.

[5] Y. Qian, H. Sharif, D. Tipper, A survey on cyber security for smart grid communications, IEEE Commun. Surv. Tutorial. 14 (4) (2012) 998–1010. Oberheide, J., & Jahanian, F. (2010, February). When mobile is harder than fixed (and vice versa) demystifying security challenges in mobile environments. In Proceedings of the Eleventh Workshop on Mobile Computing Systems & Applications (pp. 43–48).

[6] A. Minnaar, Crackers', cyberattacks and cybersecurity vulnerabilities: the difficulties in combatting the'new'cybercriminals, Acta Criminol.: African J. Criminol. Victimol. (2014) 127–144, 2014(Special Edition 2).

[7] S.L. Vrhovec, Challenges of mobile device use in healthcare, in: 2016 39th International Convention on Information and Communication Technology, Electronics and Microelectronics (MIPRO), IEEE, May 2016, pp. 1393–1396.

[8] S.A. Alabady, F. Al-Turjman, S. Din, A novel security model for cooperative virtual networks in the IoT era, Int. J. Parallel Program. (2018) 1–16.

Further reading

[1] F. Al-Turjman, Intelligence and security in big 5G-oriented IoNT: an overview, Future Generat. Comput. Syst. 102 (2020) 357–368.

[2] F. Al-Turjman, Impact of user's habits on smartphones' sensors: an overview, in: 2016 HONET-ICT, IEEE, October 2016, pp. 70–74.

[3] F. Al-Turjman, H. Zahmatkesh, R. Shahroze, An overview of security and privacy in smart cities' IoT communications, Trans. Emerg. Telecommun. Technol. (2019) e3677, https://doi.org/10.1002/ett.3677.

[4] F. Ullah, H. Naeem, S. Jabbar, S. Khalid, M.A. Latif, F. Al-Turjman, L. Mostarda, Cyber security threats detection in Internet of Things using deep learning approach, IEEE Access 7 (2019) 124379–124389.

[5] K. Kauthamy, N. Ashrafi, J.P. Kuilboer, Mobile devices and cyber security-an exploratory study on user's response to cyber security challenges, Int. Conf. Web Inform. Syst. Technol. 2 (April 2017) 306–311 (scitepress).

[6] L. Zhang, Mobile security threats and issues-A broad overview of mobile device security, in: Proceedings of the International Conference on Security and Management (SAM)(p. 1). The Steering Committee of the World Congress in Computer Science, Computer Engineering and Applied Computing (WorldComp), 2011.

[7] S. Grzonkowski, A. Mosquera, L. Aouad, D. Morss, Smartphone Security: an overview of emerging threats, IEEE Consumer Electron. Mag. 3 (4) (2014) 40–44.

[8] G. Delac, M. Silic, J. Krolo, Emerging security threats for mobile platforms, in: 2011 Proceedings of the 34th International Convention MIPRO, IEEE, May 2011, pp. 1468–1473.

[9] K. Sharma, M.K. Ghose, Wireless sensor networks: an overview on its security threats, IJCA, Special Iss. "Mobile Ad-hoc Net." MANETs (2010) 42–45.

[10] G. Kambourakis, F. Gomez Marmol, G. Wang, Security and Privacy in Wireless and Mobile Networks, 2018.

[11] A. Michalska, A. Poniszewska-Maranda, Security risks and their prevention capabilities in mobile application development, Inf. Syst. Manag. 4 (2015).

[12] D. He, S. Chan, M. Guizani, Mobile application security: malware threats and defenses, IEEE Wire. Commun. 22 (1) (2015) 138–144.

[13] N. Penning, M. Hoffman, J. Nikolai, Y. Wang, Mobile malware security challeges and cloud-based detection, in: 2014 International Conference on Collaboration Technologies and Systems (CTS), IEEE, May 2014, pp. 181–188.

[14] R. Sabillon, J. Cano, V. Cavaller Reyes, J. Serra Ruiz, Cybercrime and cybercriminals: a comprehensive study, Int. J. Comput. Net. Commun. Secur. 4 (6) (2016), 2016.

Influence of social information networks and their propagation

4

B. Raja Koti, G.V.S. Raj Kumar, K. Naveen Kumar, Y. Srinivas

Department of Computer Science and Engineering, GITAM Institute of Technology, GITAM (Deemed to be University), Visakhapatnam, Andhra Pradesh, India

1. Social influence: a brief outline

Social networks have been examined widely by social researchers for quite a long time [2,4,45]. These are to be known as tiny datasets. With the continuous appearance of informal communication destinations empowered by the internet, like Facebook, LinkedIn, and Tumblr, research on social networking has become an exceptional development because of the accessibility of free community information on an enormous scale. Social influence has prompted the improvement of usage of informal online organizations to lead to an ensuing investigation of many research questions. A productive group of such researchers has come to study the impact and data spread in informal organizations.

In this chapter, it is our point to lay out some key ideas, advancements, and accomplishments. We examine the driving applications that underlie this examination and feature significant difficulties that stay open. For accommodation and consistency of phrasing, we utilize the terms social impact examination or impact investigation to demonstrate the investigation of the dispersion of data or impact through an informal organization.

1.1 Brief introduction to social networking

With the expansion of the number of clients, data sharing over the globe has been expanded massively. To transmit the information, clients have begun making groups of members and consequently sharing information by utilizing these gatherings. These groups of clients have arranged for long-range social networking gatherings online.

The most crucial task in internet-based life is recognizing intimate gatherings of people in social organization media. Among the sites of social networking, classmates.com (1995) is considered to be the essential site that was utilized for associating clients. It was planned principally to interface present companions and past mates.

Security in IoT Social Networks. https://doi.org/10.1016/B978-0-12-821599-9.00004-2
Copyright © 2021 Elsevier Inc. All rights reserved.

Moreover, the primary drawback of these social networks sites is that the immediate linkage of companions. This could not be set up as each companion needed to be associated using the school, by utilizing the transitive investigation, then the ties were considered. The SixDegree.com webpage (1997) has beaten this impediment and was viewed as the first online system website. A few other locales were grown, for example, Friendster, Cyworld, Ryze, and LinkedIn; in any case, these destinations neglected to give the client profile customization. Though MySpace (2003) has given this option and turned as a leading online system website, this could not give the office a sharing site that also has sound substance (Fig. 4.1).

Flicker, YouTube, and Zoomr were created for sharing media substance. These products have picked up notoriety because of their client-driven nature. The principle task in internet-based life recognizes similar gatherings of people in social network media. Most of the time, to address this issue, link analysis is considered the way out. Utilizing link examination, one can discover the relationship of related clients, connections among clients in online networking.

Individual inquiries are to be comprehended, taking into consideration this link analysis, utilizing a chart that incorporates the center, a cooperating single/group, recognizing the person. Who has more companions or has regular companions distinguishes people that have the same manner of thinking. To counter these inquiries,

FIGURE 4.1

Social networks sites.

specific measures are fundamental for recognizing the center has driven gathering, evaluating the power of association and the most similar conferences. It is a measure to understand the implication of a specific audience and quantify centrality. This utilizes the indegree and outdegree of the hubs and chooses the connection between centers. If the diagram is coordinated and if the chart is undirected, we join the indegree and outdegree. However, this measure is not taken into consideration with the correlation of the hubs over the systems, e.g., Facebook and Twitter.

2. Social media "friends"

The fast advancement of data and correspondence innovation has expanded the extensive usage of online social organizations in our livelihood. Admittedly, online interpersonal organizations, for example, Sina Microblog, Twitter, and Facebook, have turned out to be a fundamental portion of our life. We access our websites multiple times to view and exchange information. These social networks have important features: they are quick, easy, and intuitive. Take Sina and Microblog, for instance, in contrast to the popular blog, which permit the utilization of cell phones to share data up to a span of 140 characters. Online social media research is essential in the context of trends from social media and viral events to political events. In the future, online social networks, as a stage for the experimental study of data, shall be a broader concern. Despite the advances that have been made, the exact investigation of data engendering is still in its earliest stages. Concentrations toward this path have mostly been slowed down by the inadequacy of accessible enormous-scale information. However, the accessibility of immense details from the online social organizations has provided phenomenal chances to probe the effect of social practices on the data spread.

Initially, data spread, in an informal online community, is dictated by rhythms and movement examples of humans. Expanding the number of late estimations demonstrates that human action designs are heterogeneous and bursty, just considering the time interim between occasions. These social movement designs are regularly depicted by a power law interevent time dispersion $P(\tau) \sim \tau - \alpha$, where τ is the time interim between two back-to-back exercises. Of late, analysts started to understand that the bursty human conduct importantly affects the spread of data (Fig. 4.2).

Furthermore, an extensive appropriation of an online interpersonal organization has expanded the challenge among data for our constrained consideration. Consistently, we get a ton of data from different online informal organizations for which we need more time and thoughtfulness regarding disbursement of each message. There is a brief question about whether such a challenge could affect the speed of social media. The issue of limiting considerations was considered through the message sent and posted on the proposed social media network. However, the limitation of concerns affecting the speed of advertising is still cloudy. In this chapter, we suggest an all-inclusive approach that includes susceptible infected, add-ons, integrated mobile applications, and limited analytics because, at that time, we had a wealth of real-life data for testing the model. The main findings of this study are expanded as follows.

FIGURE 4.2

Connecting friends: using social media.

(1) Based on observational outcomes, finding the congregation point, intuition keeps the power law appropriation with the incline ≈ 2.5. Furthermore, the dissemination of recently contaminated individuals (ascertained by the quantity of new sending every day) continues the power law with slope ≈ 1.5. Two slants fulfill the relationship of 2.5−1.5 ≈ 1.0.

(2) Through both the hypothetical research and reenactment, we demonstrate the following:

(a) If the time circulation adheres to control law with type β, at that point the rot of proliferation speed can be described by a similar force law conveyance;

(b) If bursty human conduct observes a force law appropriation with an example α, the rot of spread speed additionally keeps a force law with type $\beta \approx \alpha - 1$. In synopsis, endeavors have been made for the exploration of data proliferation. Further investigation depending on human elements is yet expected to uncover the roles of social behaviors for the data spread in an online interpersonal organization. In future inquiries, again, we can utilize other increasingly developing hypotheses to inquire about spreading elements, for example, in references [16,17].

3. Homophily or influence?

One of the significant perceptions made by social researchers is a propensity in social gatherings. Comparative individuals are associated together (after all, similar people are attracted to each other). It significantly affects the worth we get from web-based life (as frequently we hear comparative voices and communicate with similar individuals). This marvel is called homophily, i.e., love of the equivalent (Fig. 4.3) [31].

Homophily can be virtual universe scientific strategies. H. Bisgin [10] demonstrated that with massive online role-playing games. In general, it is cooperation with different players of approximate age, understanding those who live close to them in reality. Homophily held over a wide range of communications, from questing together to exchanging the in-game sales management firm. The primary way they searched for homophily was not discovered in the gender category, and they came to an idea that 32% of individuals play the game with a sentimental accomplice.

Homophily has a prescient force in online networking, to such an extent that scientists are taking a gander at last, anticipating genuine fellowships by looking at online cooperation, shared interests, and area [11]. The actual thing is homophily is so incredible, it is a rule for the entire network. On Facebook, it can be demonstrated by extrapolating from as meager as 20% of the general population [12]. Homophily has positive ramifications for protection and obscurity, as just knowing your place in a system may permit examination devices, allowing you to make assumptions about your private data with high exactness.

In Twitter, De Choudhury (2011) has demonstrated that various sorts of homophily hold multiple kinds of clients. For illustration, ordinary clients with generally a similar number of adherents followed have area and notion homophily. And they

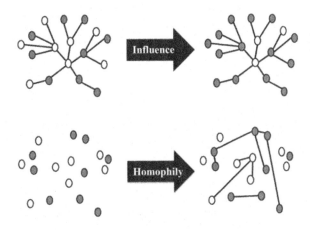

FIGURE 4.3

Homophily and influence.

will, in general, live and work close to one another and show comparable responses and perspectives. Homophily is a genuine case, a current social hypothesis. It would now be able to be investigated and be confirmed in a wide range of systems because the information is held carefully.

4. Social networks and influence

Social influence happens when others influence an individual's feelings, conclusions, or practices. Social impact takes numerous structures that can be found in similarity, peer pressure, initiative, influence, deals, and promotions. In 1958, Harvard clinician Herbert Kelman recognized three expansive assortments of social impact:

1. consistency: when individuals seem to concur with others, however, keeping their contradicting approach private;
2. the recognizable proof: when individuals are influenced by somebody who is enjoyed and regarded, for example, a renowned celebrity name;
3. disguise: when individuals acknowledge a conviction or conduct and concur both freely and secretly.

Morton Deutsch and Harold Gerard portrayed two mental needs that lead people to comply with the desires of other people. These incorporate our necessity to be correct (enlightening social impact) and our obligation is preferred (regularizing social implications). Educational result (or social confirmation) is an impact of acknowledging data from another as proof about the real world. The instructive effect becomes possibly an essential factor when individuals are unsure, because upgrades are characteristic because there is a social contradiction. Regularizing impact is an impact of adjustment to the uplifting desires for other people.

Regarding Kelman, regularizing impact prompts open consistency, while enlightening influence prompts private acknowledgment. Homophily is the inclination of people to partner and bond with comparable others, as in the adage "together" The nearness of homophily has been found in a massive swath of system considerations. More than a hundred examinations have watched homophily in some structure or another and determined that closeness breeds association. These incorporate age, sex, class, and authoritative job. People in homophilic connections share essential attributes (convictions, values, instruction, and other) that make correspondence and relationship arrangements simpler. Something contrary to homophily is heterophily or intermixing. Homophily is a measurement considered in the field of social organization. Researchers generally call it assertiveness.

Considering the probability that your friend can appear too much may be too big. It is an important finding of Christakis and Fowler [6], who argued that interactions were spread through social network sites. The examination was widely studied by the media and hailed as "one of the most powerful sources of research in restoring human science" by the National Institute on Aging [33]. Similar research maintains that social relationships were not fully separated from home. Homophily is the idea that people who similarly share things are bound to be friends. Along these lines, people are expected to act in the same way as they have the same personality. In this way, when a participant examines the bands of friends, they move in the same way, and it is difficult to determine whether the cause is social or unintentional. Even more so, consistency is always hard when people become friends with those with the wrong values.

Coming back to the bulkiness study, companions could become fat together, not in light of human virus impacts. But the fundamental comparative characteristics that drove them to be companions expanded their chances of getting stout. It is not to say that social relations do not cause firmness; however, that much work is expected to be done solely on social factors. Social impact possibly influences well-being results as well as numerous zones of necessary customer leadership. McKinsey and Company evaluated that 33% of all shopper spending, c. $940 billion in yearly utilization in the United States and Europe alone, is affected by social connections [34]. The evildoing of social impact in utilization choices makes it an appealing advertising open door for firms. To isolate social impact from homophily is alarming for the two firms and scientists. To be viable in interpersonal organizations, advertisers need to take activities compatible with the adequate procedure that aides necessary customer leadership among gatherings of associated people.

Social impact has a significant effect on numerous parts of essential client leadership and showcasing, including the reception of new advances, free community use and appropriation [24,39], e-commerce [8], new medicine treatments [10,11], the relevance of advertising [3], collaborative choice-making [15], and customer adherence [31,33]. The magnitude of impact varies with the exact situation; for example [14], finds that 35% of consumer respect is incapable of peer effects, while [18] found that one in five clients isolate their friends. Treating the effect of these investigations infers that friend impact on sociology is explorative. A progression of prominent studies recommends therapeutic and human conditions. For example, stoutness, smoking cessation, satisfaction, and depression that are generally comprehended to be noninfectious are entirely transmitted along with arranging ties [19−23].

The far-reaching infiltration of cell phones provides increasing clarity with which customers switch from PCs to portable devices and is even more pronounced with versatile applications on mobile phones. Versatile apps make up 82% of mobile phone time relative to apps. Still, four out of every five customers are considered by mobile phone ads [25]. Due to these discoveries, numerous organizations want to draw in and associate with purchasers through versatile applications. Versatile applications are frequently social, particularly those that are connected to an informal community stage. It is reasonable to imagine that friend impact assumes a job in item reception. If the distinguishing peer impact in portable application selection is not clear, the social impact can be questioned with homophily. However, we see that when one companion utilizes an application, another introduces another form. Then the question arises "is it their inactive closeness that drives selection or is it in light of friend impact?"

5. Social influence and viral cascades

Many social networks have increased, and research on this network has gained traction over the past decade. It is in part due to the incredible growth of social media websites and social networks and the continued existence of a large number of social media platforms. It is believed that people are affected when users see their social contacts performing a particular activity, so they may decide to take action as well. However, in reality, consumers show the same influence when they see it in some other social links. They may have more reasons to do so as they would have heard about it outside the online social network. Then the media may have decided that action was worth doing or may be very popular. They could be affected by seeing their affiliates perform that function. The idea behind a viral ban is to target the most influential users on the network. We can implement a chain of control response, so with minimal marketing costs, we can reach a significant portion of the system. Selecting these critical users in a complete graph is an interesting learning activity.

The current theme focuses on the analysis of detailed information sharing, motivation, innovation, and trends through communication. The fundamental question that might arise is how do we validate the models created using real data sets? Which one consists of a shared network and its broadcast about the previously revealed discoveries? Information dissemination models track various applications in the field of viral marketing: searching for rebellion, finding key blog posts to read to capture important stories, tracking leads or trends, ranking information flows and more.

The researchers studied several algorithmic problems that appear in these applications under the concept of maximizing impact. This idea can be illustrated by an example: Suppose we are offered a social network whose nodes are users. The links represent social relationships between users, in graph and impact estimates, which are between the individuals connected in the network to push output new in the market. Examining two issues with the greatest impact shows how to select a set of original users so that it ultimately affects the maximum number of users in the social network. Several fixed distribution models, including the exponential inverse model and the linear threshold model, are also based on multiplication angles. The maximum impact in classical distribution models, including both exponential cascade and linear threshold models, is measurable in terms of imaging due to several probabilistic algorithms and scalable heuristics. Various issues remain open, both from a technical point of view and in terms of the transfer of products in business applications (Fig. 4.4)

We realize that everybody buys a product, keeping in mind their condition. Incidental life occasions, impacts, and environment can additionally change our conduct. Internet-based life now profoundly impacts our shopping, connections, and training.

With some exceptions, investigations recommend that most interpersonal organizations mainly support previous social relations. Generally, Facebook is utilized to keep up existing loose connections or set free associations rather than meeting

FIGURE 4.4

Propagation based on applications.

new individuals. These connections might be delicate ties, yet ordinarily, there is some essential disconnected component among people who connect, for example, a standard class at school. It is one of the central measurements that separate online networking from prior types of open correspondence, for example, newsgroups. Research in this vein has examined how online cooperations interface with disconnected ones. Facebook clients take part in "looking" for individuals with whom they have a disconnected association more than they "peruse" for a completely unfamiliar person to meet.

Social networks are intended to be open for inhomogeneous populaces. So it is not phenomenal to discover bunches utilizing destinations to isolate themselves by nationality, age, instructive level, or different variables, regardless, whether or not that was not the expectation of the designers.

5.1 The connection between social organizations and social pandemics

Of late, a book entitled *"Connected: The Surprising Power of Our Social Networks and How They Shape Our Lives"* by Nicholas Christakis and James Fowler described how connections among people and their systems have an impact on their lives, either directly or indirectly. In their record of the particular and unusual characteristics of informal communities, the creators clarify why heftiness is infectious, why the rich get more extravagant, and even how we discover and pick our accomplices.

We like to imagine that we are, to a great extent, in charge of our everyday lives. Yet, a large portion of what we do, from what we eat to whom we lay down with, and even how we feel, is altogether impacted by people around us. Our activities can change the practices, the convictions, and the essential strength of individuals we have never met.

Social network sites can harbor a progression of significant unfortunate things such as outrage, misery, and grief, as well as useful things like bliss, love, philanthropy, and essential data. "It is the spread of the beneficial things that vindicates the entire explanation of how we live our lives in systems. If it were constantly savage to you, the connections to me and the system would crumble. Profoundly and principally, systems are associated with goodness, and goodness is required for systems to develop and spread.

Christakis's first paper on stoutness was distributed in the New England Journal of Medicine in 2007. On the off chance that somebody on the Framingham study turned out to be clinically fat, their companions were 57% more probable additionally to get large. A companion of that stout individual was about 20% bound to get large, and this was the situation regardless of whether the heaviness of the connecting companion stayed unaltered. After a year came their paper on smoking, which contained comparably capturing ties. On the off chance that an individual started smoking, the odds of their companion doing likewise expanded by 36%.

The idea that one's conduct and activities can impact associations, a stage expelled, is pretty marvelous to consider. Also, our behavior, events, and propensities are probably going to generally have more impact and be affected by online networking than we ever could have envisioned.

While thinking about the current proposals and research considering other factors, do we accept our social network? It does not feel that we become fat because of our companion; nevertheless, we do trust. It can assume a specific moment in our wellbeing in an extremely microscopic way. Moreover, the consolidated impact can roll out for an enormous improvement.

What we post in our online status has a more significant impact on our population than we might suspect. From an advertising point of view, think about how to impact previously existing clients, customers, or brand advocates on the web. Even one person can effectively make positive feelings around the crusade or brand in any capacity. The gradually expanding influence can be more unavoidable and compelling than what one may suspect.

6. Viral marketing and its impact

In the last decade, there has been a significant shift from the mainstream media. The second era of internet-based applications (e.g., "Web 2.0") (Fig. 4.5) is what is [37] proposed in the fourth version. Here population control and control letters are done in a solemn way to completely and entirely target advertising. These are viral exposure efforts [40]. This new feature opens the door to building relationships, sharing

FIGURE 4.5

Web 2.0.

them with advertisers and their clients [13] as well. Ongoing investigations demonstrate that the commercial appropriation of web-based social networking by the quickest developing US organizations is currently at a record pace (Barnes 2008). However, due to the curiosity and potential adequacy of Web 2.0, a few advertisers might be allured to depend on web-based networking media in limited time plans. Does this talk about whether online networking can reliably create viable viral word-of-mouth marketing for brands and items?

Viral marketing [38], otherwise referred to as Voice-Of-Mouth (VOM) or "buzz promotion," is a way of creating systems where smart people can trade with one another [36]. In this period of client-generated media, web-based life is not only an advertising channel: it encourages VOM. While Web 2.0 media releases content and holds open doors for advertisers, it carries with it a prospective. It irritates the need for controlling over displaying information. Before building an online brand strategy, marketers need to ask themselves how can we take inside customers to elevate items to clear missions in a reliable, controlled, and economic manner? Product-based life has given consumers their voice, not respondents as they have been in the past in partnering with the brakes. However, influential people from the brand network are sure to come into the "space" of the brand. Who works with driving profits in the life of the internet promotes a single process that can be "integrated" where advertisers encourage customers to succeed in making a product [30] Consequently, to create discomfort among customers, few advertisers have chosen to stop talking or selling "on" them, and instead, they show up with them.

Few advertisers use this technique within a corporate promotion. Instead of focusing on value-based displays, the point is to build long-term associations with clients, create a belief among consumers and sellers, so commitment is created [14]. Online relationship promotion requires the help of the procedures of association, correspondence, discourse, and worth [13]. Raising web-based life tools for showcasing incorporates ongoing video preparation and online classes that can furnish advertisers with applications. They were increasingly reliable with the idea of opening the relation to discourse. Social advertising advances additionally licensed advertisers to redo their messages and have a conversation with clients (Fig. 4.6).

Besides, the technology basis of online correspondence regularly empowers better focusing on potential clients as the databases driving locales. For example, Facebook can fragment crowds by factors, for instance, socioeconomic status and welfare, and still outline the rise of online networks as goals [8]. Nevertheless, the achievement of this standard is dependent upon significant assets being dispensed to their appropriate use and assessment. There are high and well-known aspirations that running social apps can bring about incredibly compelling advertising. For example, [7] concluded that 70% of customers visited a life-based website to obtain data. And 49% of these customers opted to make purchases with this information they found. Then 60% of survey respondents said they were likely to upload the information they find online. Other research shows that even though 90% of VOM conversations were similar, the results that appear on the internet or

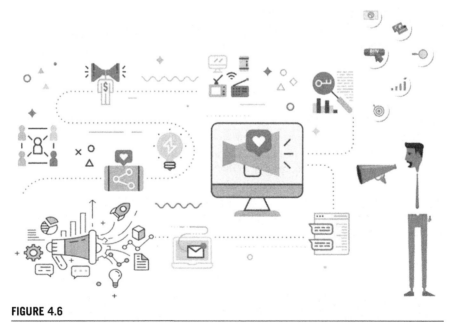

FIGURE 4.6

Social media marketing.

Source: uihere.com

social networks should not add to the fact [18]. That is visualized from eye-to-eye contact elsewhere over the phone, just 7% happen to be on the web along these lines, where VOM's positive force stays outside Web2.0 and most discussions despite everything occur. While individuals use many interpersonal communication sites and applications, they do not use personal communication goals to purchase items. They are not prone to buy things from organizations promoting such local locations [21]. Accomplishments in promoting viral strains after accepting the idea of sharing information, influencing the beneficiary responses in online systems, should be considered seriously [37]. Online influencers should be seen as learned partners in an interpersonal organization instead of being pure "advertiser" specialists. Plans that cause explicit endeavors to pick clients to advance items and administrations may disturb the norm and decrease. The adequacy of the promotive way is to deal with the hindrance between advertisers and clients, who may have yielded profits by information sharing [38]. The most important issue facing advertisers is the reliability of online entertainment resources. Customer analysis executives often assess the ratings of complete aliens, and regular persecutors know about it. In promoting efforts, a distinction should be made between unique validators [4].

A regular irresponsible expectation of social promotion is whether the efforts of online viral advertising are reflected in the short and long span. Viral advertising is feasible and measurable enough. The road to reliable measurements means that several advertisers avoid running viral web strategies. That only attracts transient consideration (for example, viral recordings) to policies that take into account prospect identification and catching of social information [6]. A lot of what occurs in social showcasing is minimal more than trial, or necessarily about "insights" as opposed to measurements. Many advertisers want to "correct" the live stream of online content and show how the front lines are, while being key drivers of the battle area in the traditional media. There is a need to influence change in theory through a "much better" approach. While a large number of social advertisers focus on site traffic, clicks, time spent on the line, and posting, dynamic social presentations persist regularly. Reliance is on a series of steps to highlight sound, quality, and client benefits for collaboration [1]. These may include one-of-a-kind guests, connection rates, applicable moves made, discussion size, discussion thickness, creator validity, content freshness and pertinence, crowd profiles, and one-of-a-kind client reach [7]. Such measurements measure whether individuals are locked in. Be that as it may, such measurements are regularly done for singular crusades and need to be considered in the predispatch stage, undeniably consolidated in message testing.

Internet life on Web 2.0 is perhaps an unusual way to discover, connect with, and produce brand advocates for key influential buyers. The trust must be built and strengthened to overcome any hesitation regarding the prospective customer. It means moving on "outdated" ways to deal with site promotion to understand relationship advertising standards. Then, create virtual conditions in which clients can connect and share experiences and basic data. One strategy for progressing logos is to move away from the solid supply and understand the idea of "cocreation."

By taking advantage of or making their online interpersonal organizations, web-based life advertisers can impact brand network and possibly influence buyer conduct. To profit by accessible chances, advertisers need to discover or set up functional brand networks and hear them out. Afterward, make unique projects and gadgets that enable potential and existing network individuals, remunerating existing shoppers to inspiring social change from potential customers. It may stimulate operating potential in a world of digital (and other audio and visual) media developed through the acquisition of high-end models of dedicated cocreators.

6.1 Various models and validation

6.1.1 Different impact models in social networks

Impact investigation is critical in social organizations [46]. Data dissemination dependent on impact is isolated into three classifications (singular impact, network impact, and impact expansion) [25]. One can comprehend the method of data dissemination through influence inquiry. The accompanying portrays the investigation and correlation of significant writing.

6.1.2 Singular impact

Singular impact alludes to assessment pioneers of related research. Estimation pioneers are the hubs who can assume a job as an extension of data dissemination. They impact different clients in a social networking organization. The impact of conclusion pioneers cannot be overlooked in data dissemination [6]. The consideration of the conclusion pioneers incorporates techniques dependent on arranging structure, current data, and client characteristics [48]. It fundamentally utilizes centrality and auxiliary openings to gauge the significance of the hubs for the primary technique. Page rank and different calculations are likewise used to rank the centers. This strategy is straightforward. However, the exactness is not high. The subsequent technique centers on shared data trade between clients. Its outcome is more goal oriented and precise than the initial; however, it is hard to use in colossal-scale information handling. The third strategy depends on clients' practices, exercises, or different variables. Even though this technique is progressively emotional, a singular impact must be looked into.

Wang Chenxu [43] proposed a strategy for demonstrating and estimating the impact of miniature-scale blog sentiment dependent on data transmission. This technique depends on the organizational structure as it was. It mines the supposition head by discovering the tipping point hub in the data dispersion process. The dynamic direct diagram portrays the procedure of data dissemination. It shows that data scattering is pitifully connected with the number of conclusion pioneers. The underlying impact of conclusion pioneers has decidedly corresponded with the name of their fans. This model could be utilized to foresee well-known data effectively.

Borgs [3] proposed a technique for discovering supposition pioneers dependent on a client's conduct and collaborations between various clients. As indicated by the competency model of the board, the clients are in an informal organization. It can be partitioned into four classes, i.e., regular clients, dynamic clients, subject conclusion pioneers, and system pioneers. The supposition chiefs can be found by utilizing both the predominant and specific elements. The client's practices, data examination, and the communication connections between clients are referred. In the end, the system sentiment pioneers are capable of attaining the three stratums of transmission. The impact of data transmission can be amplified through the mining of quantitative pioneers. Mao Jiaxin [28] proposed a technique for estimating social impact by anticipating a client's capacity to disperse data. The impact assessment depends on the retweet, the transient circulation of character retweet practices, and the time legitimacy of a snippet of data from the tweeter, and the slope of client flows is divided. This technique depends on system structure and client practice.

Wu Xianhui [47,48] projected a method reliant on the evaluation of topic leader. They connected to the weight of the charge (movement of clients, the connection between two charges and a point), the marginal weight, the quality of the client's association, and the substance attributed to the excavation of the leading evaluators for a meticulous topic.

Ullah [41,42] projected a viable model to find compelling nodes, expand data dissemination, and reduce disease instance. This model considers collaboration between centers and the topologic arrangement of the system. It places the hubs, in the beginning, dependent on the load between the different centers.

The impact between clients is chosen by the fleeting associations of the client and its neighbors. Secondly, impact centers with the highest degree of K are selected by neighbors and topologic associations. A large portion of the examinations in singular impact investigates a center around mining supposition pioneers. As a rule, if you need to discover the feeling heads, you should know the most affected clients in an informal organization. The correlations of the individual impact techniques are listed in Table 4.1. One can think about them from three angles: structure organization, client associations, and client characteristics. These are the fundamental strategies for a singular impact look. There can be just a single component included, e.g., tipping point hub. It can likewise incorporate a few mixes of ingredients, e.g., client action and centrality [44], action and access time dissemination, movement, and connections. A client's response is significant for a singular impact to inquire about; obviously, utilizing a blend of these three components gives better outcomes and creates increasingly exact data. Despite the strategies being used, all require a quantitative basis with which to weigh the impact. The base can be out level from either a hub, campaigner or middle person, the capacity of dispersion, inclusion, and cooperation. These techniques consistently start with a harsh determination dependent on the system structure for feeling pioneers first. At that point, the communications and client ascribed are abused to make the ultimate result progressively. Cooperation-based individual impacts have been given more consideration lately.

Table 4.1 Comparison of the singular impact methods [46].

Researcher	Network structure	User communications	User characteristics			Method	Quantitative decisive factor	Appliances
			User behaviors	Other features				
Wang Chenxu [43]	YES	NO	NO	NO		Community group investigation	Outdegree	Identify opinion leaders and prediction
Borgs [3]	NO	YES	YES	Centrality		Capability	Advocator, centrality, and mediator	Recognize estimation influential and authority maximization
Mao Jiaxin [28]	YES	NO	YES	Entrée time		Community group investigation	Potential of dissemination	Authority predicting
Wu Xianhui [47,48]	YES	YES	YES	Topic and weight		Page rank	Coverage and correction	Mining topic opinion leader
Ullah [41,42]	YES	YES	YES	Neighbors of neighbors		Community group investigation	Advocator	Recognize significant nodes

6.1.3 Community impact

A people group is a gathering of individuals with some natural properties. In social organizations, people consider different frame networks based on interests. A group of people is a subset of a system in which clients are strongly associated and have comparable characteristics, such as playing badminton, or their investigation zone is parallel. Even though the configuration of community associations changes due to time, networks tend to stay firm. The first test is how to identify the well-built influence networks that are within the communal organization. For this purpose, several approaches have been worked out, consisting mainly of links and qualitative features. Previous research could differentiate networks according to the social bonding method, in which content center characteristics [49,50] as a model [51] projected a PCL-DC technique dependent on a discriminatory probability model. This technique is used to quantify networks associated with attachments and substances. The likelihood of joining a couple of nodes is not merely rendered by familiarity, but also by content. It uses an expectation maximization calculation to enhance network participation probabilities and substance loads. At last, every hub is disseminated to a network with the highest likelihood.

Zhou [52] designed the SA-Cluster-Inc technique that identified right paths from a hub to another by inserting implicit hubs and boundaries in a new quality diagram. Furthermore, K-means bunch calculations are used to group the first charges. As the grouping pattern suggests, the local arbitrary walkout grid is updated on each cycle. Accordingly, the enlargement framework is determined rather than a full grid estimate in the SA-segment model, making this model more and more productive.

Ruan [35] further enhanced the perspective of drug-related connections; his work suggests, however, that this technique is not as professional as it could be. They offered an effective CODICIL technique for network recognition by joining and linking. Connection quality is selected by the probability that the relationship remains inside the system. The probability of an element is evaluated by the cosine or Jasson's coefficient. First of all, it creates layers of fabric. Second, the yarns and their weights are used together to get along. Over time, the practical benefits of adjacent mapping sites are obtained using a one-sided analysis, and the network is divided by calculating the Metis and Markov calculations. Yang [50] agreed with the views of the aforementioned critics but decided to give the text the "characteristics" and suggested cities from edge structure and node attributes.

Yang and Manandhar [49] consider that substance- or character-related links are not a perfect technique for discovering evaluation-based networks. They planned a method that consolidates associations, themes, and assessments to identify various systems with an alternate subject conveyance that clarifies the structure of cover networks. From the slanted point of view, this strategy has specific agent importance. Peng [32] believes that there are 10 associations in any event for every hub in a network. Although the number is significantly lower than the number displayed in the graph view, it can speak of the structure of the diagram. First, they receive a K-center calculation to separate the K-center subdiagram. Networks are identified using network discovery calculations and updated calculations. Gurini et al. [9]

believe that an opinion-based strategy is not objective. They proposed a JRC technique strategy that reflects on the actual behavior of the client and on the volume and objectivity of the related substance that has been created. In this technique, $q = 0.8$ speaks to the proximity boundary. An edge is formed when the respect between two nodes is more remarkable than q. The calculation of bunches consists of two stages: the first step is to locate the individual parcel of the system. The subsequent progress is to establish a worldwide limit of measured quality when the inner circles are merged into two assemblies.

These two steps are repeated to identify closed-loop nets. Ullah [42] introduced a model of cognitive processes in terms of reliability and similarity. Confidence between two sites in an informal organization is influenced by two thresholds between the source hub and the target site. Confidence management provides guidance. The task of universal comparisons is incredibly complex. In this example, the first sections are designated as public places, as indicated by the first online numbers that use the weight gain by the trust. At that point, hubs are allocated to the systems dependent on a comparability limit. Network identification is the premise of the network impact.

From the previous examination, one can see that exploration is network recognition-based substance or traits based and notion-based techniques. In any case, the method is dependent on joining, which is not precise and not fit for the investigation of dynamic informal communities. An important property or substance has affected the network identification process. In this way, most specialists use a mixture of the two substances, and the group-based strategy is temporarily applied. In group strategies, the first step is to build a system structure dependent on the substance or properties. Second, the underlying structure is restored by an iterative link. The primary goal is to improve network recognition accuracy and reduce time usage. Different properties are considered for accuracy. Written research shows that the model exhibition may not improve if we consider the excessive number of characteristics. For example, an investigative inquiry would not be appropriate to identify all networks or would not be useful to specific entities or circumstances. A correlation of network location calculations is introduced in Table 4.2.

6.1.4 Influence maximization

The maximizing impact is at the heart of social organizations. The idea of impact expansion was initially proposed by Kempe [19] with minimum productivity. To extend the reach, numerous analysts have conducted many follow-up examinations, such as the IRIE model and the IPA model. Even though the productivity of these models is superior, they are not thus far precise enough. Borgs [3] devised an inverted impact assessment technique to improve accuracy; however, the time of the inspection is greater. Tang [39] projected strategies that promise the exactitude of the model; nevertheless, the time essential is excessive. As indicated by the examination up to now, the first test is finding the seed hubs in impact amplification. These strategies depend on impact likelihood, eager calculations, and heuristics calculations. An investigation of critical writing is examined beneath. Lei [25] accept

Table 4.2 The judgment of the main aforementioned algorithms (Mei Li, 2017).

Model	Associations	Characteristic/contented	Opinion	Technique	Quantitative decisive factor
PCL-DC	YES	YES	NO	Likelihood	Solidity and entropy function
SA-cluster-Inc	YES	Creative and theme	NO	Bunch	Excellence function
CODICIL	YES	Stem words, label, and circumstance tag	NO	Bunch	
Hypothesis based	YES	Consumer, wording	YES	Likelihood	Sentiment topic resemblance
SVO	YES	Benefit	YES	Bunch	Homophily
Interest and trust based	YES	Significance, confidence	NO	Equally	Excellence function

that this impact likelihood is not accessible or inadequate at times. They propose a model for maximizing online impact (OIM). In this model, the impact centers are selected depending on the circular cycle strategy. The model first utilizes access to enforce data to obtain basic impact centers. Subsequently, seed splitters are chosen using the explore-exploit methodology, which depends on the impact of rivalry.

Third, as indicated by the clients' criticism, the impact data is refreshed to finish the emphasis. At last, the impact amplification hubs can be acquired through a few cycles, as indicated by the clients and the market spending plan. This model has extraordinary focal points for a decision of an item that deals when there are a few comparable items prepared by a mixture of creators. Lin [26] proposed STORM, STORM-Q, and STORM-QQ dependent on the multi-round competitive authority maximization technique. Within these models, the most compelling gathering of hubs can be acquired by great emphasis on a few system gatherings [14]. Expelling the investigation into impact hubs is restricted in a particular range. One cannot pick any impact hub in an entire informal organization for web-based transactions. Then the impact hubs can be chosen from among the clients who enclose the products as they were. However, the large purchasing customers cannot be considered the best or largest purchasers from a client. Maximizing implies discovering the impact clients in a specific range and is obscure. To take care of the issue [14], proposes a smart strategy. Clients may not be persuasive clients; however, they may now know the powerful clients that are dependent on their friendships.

Li [27] has paid good attention to reach individuals paying attention for quite a long time. During the calculation, exchange and proper consistency are used for the questions. They proposed a technique of intensification of influence, which depends on the knowledge of religion. Besides, they conducted a bit of research to increase the competitive edge. They designed the GET REAL model based on the game hypothesis. According to strategy, the bundle of the system is considered to be a candidate that has an impact on interpersonal organizations. The determination methodology, which is the best part of the system, attained discovery of the equilibrium point in each round. The predictable result is then well thought out as an income in this game [29]. It is supposed that most scientists considered medium-sized title centers but ignored fragile relationships. They designed an ideal permeation model to suppress impact amplification. In this model, the capacity of fragile relationship centers is underlined, and latent individual associations are exposed by weak links. It is clear from the previous review that impact investigation centers have an impact on both personal and network levels.

The main objective of these two levels is to find and enhance the germ nuclei. The correlation of the impact amplification techniques in writing is shown in Table 4.3. Impact spread consistently examines the information and relies on the model. In the model-based calculation, the known impact scattering model shall first be given; at this point, a specific heuristic calculation can be used to select seed fungi. It is not universal for some system topologies. In any case, the investigation of interpersonal organizations depends on accurate informal community information in information-based models. The definitive model is achieved by a method for learning progress.

Table 4.3 Assessment of authority maximization method (Mei Li, 2017).

Model	Locate beginning	Selection of beginning nodes	Information/ mold driven	Repetitive round	Recurrent modernization/ items/information	Purpose
OIM	YES	Survey utilize, heuristic	Replica	YES	NO	Individual influence maximization
Adaptive seeding	YES	Familiarity inconsistency	Data	NO	NO	
CASINO	YES	Agreement attentive is mentioned	Data	YES	NO	
Optimal percolation	YES	The implication of delicate nodes	Data	NO	NO	
STORM	YES	Maximization of the total going	Data	YES	YES	Competitive influence maximization
GETREAL	YES	Game theory	Model	NO	YES	

As a result, these models are incredibly versatile. The impact amplification model consistently embraces two parts: the selection (training phase) and the operating step (aggressive phase). Seed nozzles can be chosen using one or more spherical techniques. In the multicircle method, the intense effect is used to refurbish seed fungi in the next cycle. Empowering the individual is found continuously at a specific point or fraction of the data. Whatever it is, when it comes to aggressive research, it is regularly based to rely on various things or data. The primary target is to boost the gathering's impact on whether the adversary's procedure is known [20] or obscure.

The model has to be training based and designed to adapt to a set of conditions. There are a few advances through different cycles, making it progressively flexible and versatile. These key performance indicators are often found in the independent cascade model, the linear threshold model, and the game hypothesis model [5].

7. Case study and applications

Contextual analysis is a "genuine report identifying with a specific occasion, over some undefined time frame." They assume an essential job in the profound comprehension of an occasion and encourage the data relating to the procedure of such an opportunity. A contextual investigation via web-based networking media and showcasing brands draws out innovativeness and advancement. Among the brands, what parts have been effectively making procedures to advance their crusades through internet-based life needs to be assessed.

The last few years have had an incredible impact on the media outlook of India. Development is not limited to the use of a website. There was a considerable intrigue that appeared by many Indian brands on this stage. There is no specific request. Here is the rundown of brands whose contextual investigations have been depicted in their web journals/site/designated for grant/composed by blogger/broken down by web-based life devotees. Some of the arguments are these:

1. how Facebook applications use brands: a crusade for technologic social media deployment
2. Adidas on Facebook, Cricket social Media and Contextual Screening Ax, Facebook Marketing Case Study
3. Asian Colors, Tag your Holi Friend Facebook campaign
4. clients view on social media, HDFC Bank on
5. Kingfisher Bearup: How Kingfisher Beer folds Tweety and increases brand respect
6. newspaper tweeting: Volkswagen action based on a fair mix of conventional and social media
7. listening to social media: Snapdeal.Com; how a wrong posting leads to a terrible online acquaintance
8. Evalueserve: using LinkedIn for leadership and brand building
9. inspirational strategies through India's Volkswagen drive that motivated 2700 proposals in 4 weeks; On LinkedIn Dance India Dance: how they committed themselves to the famous case study of TV shows and social media on social media
10. corruption free India, strategies adopted by Anna Hazare

11. MTV roadies: complete reconciliation of social media and the copious dangerous brand for adolescents in the nation
12. how the Unbranded Network Works on Facebook Case Study of the Shari Photography Academy, as it turned out Nando's discount, on Peri-Peri discount appeared on Facebook
13. Channel V: you can use Twitter to make sounds about relaunching a channel [V]
14. IPL doubled the intensity of internet life with observers from around the world; YouTube context investigation
15. JustBooks Clc strategies to connect intended interest groups by producing memory games
16. 7UP Lemon Pattalam: Facebook marketing case study
17. Facebook vote on Hard Rock café
18. how Indian magazines use internet social networks; Remembers: Vogue India; Forbes India
19. it uses social media as a segment of human services in India: a Narayan Netralay case study
20. Ching's Secret: India's best known social media brand

8. Summary

Commonsense applications persuade the vast majority of the research shrouded in this book. What is more, there is a valid proposal hidden in this work: viral wonders cannot exclusively be displayed precisely. However, they can likewise build; for example, in the territory of advertising, there is a particular desire that a battle can be intended to "become a web sensation." All in all, interpreting hypothetical outcomes into training is once in a while direct, and data/impact proliferation investigations are no exception. However, we have demonstrated that experimental results show impact undoubtedly. It is a genuine marvel that looks like a portion of the deliberations that have been proposed to illustrate it. Then again, a few inquiries, despite everything, remain about how ventures can use the consequences of this exploration and effectively convey the strategies in a natural, viral-promoting setting.

We have seen incredible advances in principle, research, and techniques identified with viral marvels. As referred to in the anterior segment, there are many open research bearings, so we imagine this a fertile research territory for a considerable length of time to come. A significant update, however, is that building viral wonders will bring them to life. A substantial number of the applications talked in this book still cannot seem to be removed from the research center and set up as a regular occurrence at an enormous cultural or mechanical scale. The social network is significant without a moment's delay for the general public, the industry, and the researchers, as commonsense issues without a doubt spur a large portion of this exploration. This territory can be profited by conveying the arrangements created by the examination network in viable settings and seeing how these arrangements act practically and what new difficulties emerge there. To wrap things up, shutting the hole among hypothesis and practice is critical for the general public at large to have the option to receive the rewards of this exploration.

References

[1] R. Angel, J. Sexsmith, Social networking: the view from the C-suite, Ivey Bus. J. 73 (4) (2009), 14818248, Jul/Aug.

[2] D. Barnes, G. Nora, Society for new communications research study: exploring the link between customer care and brand reputation in the age of social media, J. New Commun. Res. Iii (1) (October 2008) 86–91.

[3] C. Borgs, M. Brautbar, J. Chayes, B. Lucier, Maximizing Social Influence in Nearly Optimal Time, arXiv, 2012. arXiv:1212.0884.

[4] B.D. Deebak, F. Al-Turjman, A Novel Community-Based Trust Aware Recommender Systems for Big Data Cloud Service Networks, Sustain. Cities Soc (2020) 102274.

[5] W. Chen, Y. Wang, S. Yang, Efficient influence maximization in social networks, 28 June–1, in: Proceedings of the ACM SIGKDD International Conference on Knowledge Discovery & Data Mining, Paris, France, July 2009, pp. 199–208.

[6] D. Deebak, F. Al-Turjman, L. Mostarda, Seamless Secure Anonymous Authentication for Cloud-Based Mobile Edge Computing, Elsevier Comput. Electr. Eng. J. (2020).

[7] T. Fisher, ROI in social media: a look at the arguments, Database Mark. & Cust. Strategy Manag. 16 (3) (2009) 189–195.

[8] P. Gillan, B B; 6/8/2009, Tap into Social Communities, vol. 94, 2009, 8, 15–15, 1/2pp.

[9] D.F. Gurini, F. Gasparetti, A. Micarelli, G. Sansonetti, Analysis of sentiment communities in online networks, in: Proceedings of the International Workshop on Social Personalisation & Search Co-located with the ACM SIGIR Conference, Santiago, Chile, 9–13 August 2015, pp. 1–3.

[10] H. Bisgin, N. Agarwal, X. Xu, Investigating homophily in online social networks, in: 2010 IEEE/WIC/ACM International Conference on Web Intelligence and Intelligent Agent Technology, Toronto, ON, 2010, pp. 533–536.

[11] H. Al-Qaheri, S. Banerjee, G. Ghosh, Evaluating the power of homophily and graph properties in social network: measuring the flow of inspiring influence using evolutionary dynamics, in: 2013 Science and Information Conference, London, 2013, pp. 294–303.

[12] R. Sakthivel, G. Nagasubramanian, F. Al-Turjman, M. Sankayya, Core-level cybersecurity assurance using cloud-based adaptive machine learning techniques for manufacturing industry, Trans. Emerging Telecommun. Technol. (2020). https://doi.org/10.1002/ett.3947.

[13] S. Harridge March, S. Quinton, Virtual snakes and ladders: social networks and the relationship marketing loyalty ladder, Market. Rev. 9 (2) (2009) 171–181.

[14] T. Horel, Y. Singer, Scalable methods for adaptively seeding a social network, in: Proceedings of the 24th International WorldWideWeb Conference (WWW2015), Florence, Italy, 18–22 May 2015, pp. 1–14.

[15] L.I. Hui, B. Shen, J. Cui, J. Ma, Ugc-driven social influence study in online microblogging sites, China Commun. 11 (2014) 141–151.

[16] J.G. Oliveira, A.L. Barabasi, Human dynamics: darwin and Einstein correspondence patterns, Nature 437 (7063) (2005) 1251.

[17] K. Jung, W. Heo, W. Chen, IRIE: Scalable and robust influence maximization in social networks, in: Proceedings of the 2012 IEEE 12th International Conference on Data Mining, Brussels, Belgium, 10–13 December 2012, pp. 918–923.

[18] E. Keller, J. Berry, Word-of-Mouth: the Real Action is Offline, 2006 (Accessed February 2, 2010) Available at: http://www.kellerfay.com/news/%20Ad%20Age%2012-4-06.pdf.

[19] D. Kempe, J. Kleinberg, É. Tardos, Maximizing the spread of influence through a social network, in: Proceedings of the 9th ACM SIGKDD International Conference on Knowledge Discovery & Data Mining, Washington, DC, USA, 24—27 August 2003, pp. 137—146.

[20] J. Kim, S.K. Kim, H. Yu, Scalable and processing of influence maximization for large-scale social networks?, in: In Proceedings of the 2013 IEEE 29th International Conference on Data Engineering, Brisbane, Australia, 8—12 April 2013, pp. 266—277.

[21] Knowledge Networks Press Release, Internet Users Turn to Social Media to Seek One Another, not Brands or Products, 2009. May 20, (Accessed February 2, 2010), Available at: http://www.knowledgenetworks.com/news/%20releases/2009/052009_social-media.html.

[22] L. Barabasi, The origin of bursts and heavy tails in human dynamics, Nature 435 (7039) (2005) 207—211.

[23] L. Weng, A. Flammini, A. Vespignani, F. Menczer, Competition among memes in a world with limited attention, Sci. Rep. 2 (2012). Article 335.

[24] S. Lei, S. Maniu, L. Mo, R. Cheng, P. Senellart, Online influence maximization, in: Proceedings of the 21st ACM SIGKDD International Conference on Knowledge Discovery and Data Mining, Sydney, NSW, Australia, 10—13 August 2015, pp. 645—654.

[25] H. Li, S.S. Bhowmick, J. Cui, Y. Gao, J. Ma, Get real: towards realistic selection of influence maximization strategies in competitive networks, in: Proceedings of the 2015 ACM SIGMOD International Conference on Management of Data, Melbourne, Australia, 31 May—4 June 2015, pp. 1525—1537.

[26] H. Li, S.S. Bhowmick, A. Sun, J. Cui, Conformity-aware influence maximization in online social networks, VLDB J. 24 (2014) 117—141.

[27] H. Li, J. Cui, J. Ma, Social influence study in online networks: a three-level review, J. Comput. Sci. Technol. 30 (2015) 184—199.

[28] J.X. Mao, Y.Q. Liu, M. Zhang, S.P. Ma, Social influence analysis for the micro-blog user based on user behavior, Chin. J. Comput. 37 (2014) 791—800.

[29] F. Morone, H.A. Makse, Influence maximization in complex networks through optimal, Nature 524 (2015) 65—68.

[30] A. Needham, Word of mouth, youth and their brands, Young Consum. 9 (1 2008) (2008) 60—62.

[31] M. McPherson, Homophily in social networks, PT - Journal Article DP, Ann. Rev. Sociol. (2001) 415—444.

[32] C. Peng, T.G. Kolda, A. Pinar, Accelerating Community Detection by Using K-Core Subgraphs, arXiv, 2014. arXiv:11403.2226.

[33] NIA (National Institute on Aging), Action Plan for Aging Research: Strategic Plan for Fiscal Years 2001—2005, 2001. Retrieved May 31, 2005, from.

[34] Q. Yan, L.L. Yi, L.R. Wu, Human dynamic model co-driven by interest and social identity in the MicroBlog community, Physica A 391 (2012) 1540—1545.

[35] Y. Ruan, D. Fury, S. Parthasarathy, Efficient community detection in large networks using content and links, in: Proceedings of the 22nd International Conference on World Wide Web, Rio de Janeiro, Brazil, 13—17 May 2012, pp. 1089—1098.

[36] A. Sheikhahmadi, M.A. Nematbakhsh, A. Zareie, Identification of influential users by neighbors in online social networks, Physica A 486 (2017) 517–534.

[37] S. Clare, C. wai, The Facebook Era: Tapping Online Social Networks to Build Better Products, Reach New Audiences and Sell More Stuff, Pearson Education, MA, 2009.

[38] M.R. Subramani, B. Rajagopalan, Knowledge-sharing and influence in online social networks via viral marketing, Commun. ACM 46 (12) (2003) 300–307.

[39] Y. Tang, X. Xiao, Y. Shi, Influence maximization: near-optimal time complexity meets practical efficiency, in: Proceedings of the 2014 ACM SIGMOD International Conference on Management of Data, Snowbird, UT, USA, 22–27 June 2014, pp. 75–86.

[40] R. Thackeray, B.L. Neiger, C.L. Hanson, J.F. McKenzie, Enhancing promotional strategies within social marketing programs: use of Web 2.0 social media, Health Promot. Pract. 9 (2008) 338.

[41] F. Ullah, S. Lee, Community clustering based on trust modeling weighted by user interests in online social networks, Chaos Solit. Fractals 103 (2017) 194–204.

[42] F. Ullah, S. Lee, Identification of influential nodes based on temporal-aware modeling of multi-hop neighbor interactions for influence spread maximization, Physica A 486 (2017) 968–985.

[43] C.X. Wang, X.H. Guan, T. Qin, Y.D. Zhou, Modelling on opinion leader's influence in microblog message propagation and its application, J. Softw. 26 (2015) 1473–1485.

[44] Y. Wang, W.J. Huang, L. Zong, T.J. Wang, D.Q. Yang, Influence maximization with limited costs in the social network, Sci. China Inf. Sci. 56 (2013) 1–14.

[45] M. Li, X. Wang, K. Gao, S. Zhang, A survey on information diffusion in online social networks: Models and methods, Information 8 (2017) 118.

[46] S. Wasserman, K. Faust, Social network analysis methods and applications, Struct. Anal. Soc. Sci. 91 (1994) 219–220.

[47] X. Wu, H. Zhang, X. Zhao, B. Li, C. Yang, Mining algorithm of microblogging opinion leaders based on the user-behavior network, Appl. Res. Comput. 32 (2015) 2678–2683.

[48] X.D. Wu, Y. Li, L. Li, Influence analysis of online social networks, Chin. J. Comput. 37 (2014) 735–752.

[49] B. Yang, S. Manandhar, Community discovery using social links and author-based sentiment topics, in: Proceedings of the 2014 IEEE/ACM International Conference on Advances in Social Networks Analysis and Mining, Beijing, China, 17–20 August 2014, pp. 580–587.

[50] J. Yang, J. Mcauley, J. Leskovec, Community detection in networks with node attributes, in: Proceedings of the 2013 IEEE 13th International Conference on Data Mining-Workshops, Dallas, TX, USA, 7–10 December 2013, pp. 1151–1156.

[51] T. Yang, R. Jin, Y. Chi, S. Zhu, Combining Link and Content for Community Detection, Springer, New York, NY, USA, 2014, pp. 190–201.

[52] Y. Zhou, H. Cheng, J.X. Yu, Clustering large attributed graphs: an efficient incremental approach, in: Proceedings of the 2010 IEEE International Conference on Data Mining, Sydney, Australia, 13–17 December 2010, pp. 689–698.

Pragmatic studies of diffusion in social networks

5

B.D. Deebak, Sanjiban Sekhar Roy, S. Kathiravan

School of Computer Science and Engineering, Vellore Institute of Technology, Vellore, Tamil Nadu, India

1. Introduction

Social networks have become the most popular tool for the use of communication and data sharing. They used networking platforms to attract the interest group that had a six-degree social networking platform in 1997. Owing to the production of smartphones, the interested users (71%) contain personalities, candidates, and profitable organizations to address their existence in social networks. They extensively use social tools to commercialize their personal use in the social market and media entertainment. They allows social users to create and ingest a huge amount of information that plays a crucial role in different chores such as epidemiologic sponsoring, administrative promotions, and career search. As of now, the prevalent social network is Facebook, which has 2.5 billion monthly active users. Because of the massive size of social networks and excessive generated data, it is extremely challenging to address the issues of diffusion and summarization of information. Therefore, these problematic issues are highly represented to highlight the difficulties of information diffusion and information summarization in social networks.

Several factors may infer the technical elements, namely, network connectivity, position, relocation timestamp, content posters, etc. Moreover, most of the common diffusion factors such as network connection, content posters, and signposts investigate the impact of information diffusion. To address the issues effectively, these network factors are extensively studied, which diffuses the information to improve the content efficiency. The time and content factors may cautiously be permeated by the information to a larger extent. To propagate the information, the factors of user activity and network connectivity are determined. To examine the activities of a confidential authority, a source of information or entertainment is chosen that tries to disseminate private information. Therefore, a novel algorithm is presented to discover the sensitive topic that may be authorized in social networks. It uses the subjective position and the authoritative users to stimulate the word-of-mouth (WoM) market in particular. It is obvious that the contents of social media are very limited to exist in the networks.

Security in IoT Social Networks. https://doi.org/10.1016/B978-0-12-821599-9.00005-4
Copyright © 2021 Elsevier Inc. All rights reserved.

Assuming that the message content is posted and is in progress of review by the targeted audience, then the content may or may not have a chance to attract more users to receive the higher audience feedback. To determine the best posting time(s), the high information diffusion may be preferred that uses the audience interaction to classify the content types of social media. Users may act in response to the types of social content. As such content is related to a specific subjective topic, then it will be more poignant or controversial to collect more remarks. Additionally, sentimental contents play a crucial factor to gather the user response. They may receive diverse reactions to rationalize the course of sentiment policies. It is understood that these sentimental policies may be used to summarize the problematic information to latch the user's attention. Lastly, the problems of diffusion and summarization are carefully observed to examine a large volume of social contents that may be difficult to search or analyze in social networks. Notably, the factor called information summarization creates a summary structure concisely to infer the large volume of unstructured information.

To address the issue, a novel method is presented that summarizes unstructured media contents to generate the subjective topics. To identify the social stimulus contents, a novel method is preferred that analyses the sentimental contents to determine whether a post has a high arousal or not.

1.1 Issue: information diffusion

Information diffusion involves a systematic process to disseminate information that entreats an individual or a community to form a centric communication in a social network [1]. The social network understands the implicit use of information diffusion to improve industry performance, spectator commitment, and custom-made recommendations that may develop a reliable mining of opinion [2]. The diffusion may typically be successful when the information extends to a large group of communities in the network. The networking services may permit the social group to spread the information out over various feedbacks such as reference, share, and media retweet. Businesses, societies, and influential people in social networks may desire to upturn the information diffusion over a large number of feedbacks.

Post content in media may increase the content visibility using a large number of feedbacks that build the status of the content maker and entice additional users to provide their feedback. The existing studies reveal that the information diffusion mainly focuses on (1) how does a piece of information disperse widely in social networks [3], and (2) how does the information diffusion enhance the data-driven model [4]. The core motivation of this chapter is to improve the diffusion factors through the factors that optimize the resulting outcome.

1.1.1 Determinant factors

An abundant volume of social content is generated periodically that obtains the objective of the content maker to propagate their facts to a large viewer. However, several key factors closely connect to information diffusion including network connection, content posters, and signposts. Therefore, this chapter widely studies three essential factors, viz., network connection, content posters, and signposts, which have cognitive impact upon information diffusion to advance original methods to raise information diffusion.

1.1.1.1 Network connection

A prominent authoritative of a social network may involve diffusing the information that extensively compares the activities of an ordinary user on the network. To discover the access of authoritative users, a social network may effectively be modeled as a graph, where connected nodes represent active users and connective edges represent a link between the active users. The prominent authoritative may advertise a subjective topic over a WoM market. The WoM market is becoming a reliable and cost-effective system that advertises market products through families, people, or recognized authorities. In this chapter, a novel algorithmic strategy is presented to discover the influential users, i.e., online social group, namely, Facebook.

As the position of prominent authority often diverges through the subjects [3], it is an additional benefit to discover sensitive subjects of authority. To this end, the sensitive subjects are discovered that form a social interface where weighted edges are dynamic and based on group interactions and similar interest of a post that advertises a topic widely [5]. The group interactions may be in the form of like, comment, and share. To detect the prominent sensitive subjects of the authority, a link analysis from social networks may be preferred that analyzes the sensitive subjects in the interaction graph, i.e., social group. The perception of the WoM secure market may be proposed to promote the subjects and to improve the market effectiveness, where multiple sensitive subjects represent the influential factors of authority. To enhance effective marketing and achieve higher visits, the preeminent time is so important.

1.1.1.2 Signposts

The generation of media content is precise that typically involves 50% of feedback receiving the positions on Facebook. The key complication is to address the high data diffusion to compete for a post over various other posts surrounded by a short lifetime. If any posted content is available, then the audience may show interest to interact with the subject contents to collect a huge amount of feedback. Therefore, a subjective post with less feedback will not spread over a large community. On the other hand, a subjective post made at a precise time may conduct a community interaction over the available quantities of feedback, and thereby the information diffusion may be enhanced. To regulate the post rank from the audience feed, a newsfeed ranking may be deployed that computes the interactive feedbacks in the social networks.

In this chapter, the best posting time(s) is effectively utilized to analyze the content types to calculate the audience's feedback. To analyze the interactive social contents, five domains of Facebook pages, namely, e-commerce, transportation, telecommunication, sanatorium, and government, are used. To discover the posting time, two scheduling classes are driven: posting-based and feedback-based. The former schedule computes the creation of the posting time, where the Facebook admins may not be responsive while they rely on the post to acquire maximum listener feedback. Conversely, few Facebook admins with the acquaintance of a newsfeed might have an intuition to rank the post, i.e., maximum listener feedback. Therefore, a posting-based schedule is proposed to maximize the listener feedback that is constructed using frequent posting timings [6,7]. Fig. 5.1 illustrates the average feedback per post over time intervals.

The audience feedback timing is chosen to examine the created posts that endorse the best time to generate a new post influencing higher audience feedback. The posting-based schedule is proposed to classify feedback schedules into three types that compute the values of the posting using feedback gain. It is observed that the feedback gain may be chosen to identify whether the feedback-based schedule is higher than the posting-based schedule. The finest feedback-based schedule may be seven times more than the amount of audience feedback linked with the regular amount of listener feedback. It would optimize the posting schedule to determine the content types that may increase the audience visit such as audio, image, videos, etc.

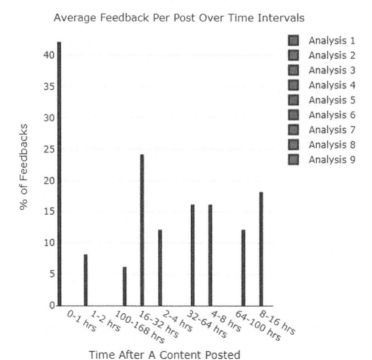

FIGURE 5.1

Average feedback per post over time intervals.

1.1.1.3 Content posters

The content poster shows a critical role to regulate the success ratio or reputation rate in social networks. Naveed et al. [8] exhibited that the poster content measures the reputation rate of listener feedback. Three common types of listener feedback in social media consider like, comments, and share to compute the informative topics. If a content post is gaining more attention, i.e., listener comments, then in social media, it will render a quality position because many individuals might have read the post and also left their consistent comments. High excitement content may be beneficial in numerous applications that involve new perception, recommendation systems, information diffusion, and collective behavioral opinions. In this chapter, a high arousal infers news posts published in social news. An unsupervised approach is preferred to mark whether the post has a high arousal or low arousal [9].

Two classes such as high and low arousal are built to extract several features that train them using ensemble-based voting classifiers. Assuming that social media has a new post, we ought to predict whether the post may generate a high arousal or not. The subjective topics of high arousal are used to determine whether it leads to high arousal or not, i.e., for a new subjective post. On the contrary, a content post with a precise sentiment may improve the information diffusion [10]. The precise sentiment is a key factor to gain a reputation that attracts the attention of the larger groups in social media. Owing to the nonexistence of intrinsic guidelines, social media is being demoralized a to range of post as more or less aggressive news. In this chapter, channels such as television, radio, and newspapers are used to analyze the sentiments to gather user responsiveness in social media.

Using feedback information, the sentiment policy can be understood that allows the user to express their opinions on a news post. Moreover, it completely depends on the sentiment of a news post to analyze the post ranking. However, the existing recommendation systems may enhance the posting contents using sentiment dynamics. Users may have their preferences of the new topics to read more exciting news that would recommend a good reputation channel to gain more sentiments.

1.2 Issue: information summarization

The social websites periodically produce a massive volume of sensitive data. Of late, 90% of web media data has been generated [11]. Facebook sporadically makes 4 petabytes of data per day using the microblogging platform. Twitter produces 500 million tweets every day. Because of a substantial amount of data, searching, comprehending, or categorizing is difficult for classifying the data. Information summarization is a procedural process that creates a short comprehensible summary from the unstructured data. It may use a summary profile of a user to generate real-world events that improve content tags, search, and categorization. The essential approach is used to discover the key topics from the massive data. The generated topics often gather text categorization that is often terminated. In this chapter, a novel method is presented to give a summary of unstructured texts. It uses natural language processing (NLP) and associative mining to classify whether the subjective topics are very similar or not to create manually [12]. A comprehensive summary is planned to relate the semantic grouping using embedded words.

1.3 Major contributions

The key contributions are as follows:

1. Providing a subjective topic, subject-sensitive WoM marketers are considered to maximize the influential factors of the information in the media network. A novel method using a top-sensitive interaction graph is proposed to discover sensitive WoM marketers.
2. To regulate the preeminent time, a new post may lead to increase of user feedback, i.e., on the content maker. The preeminent posting can contribute more listener feedback compared to the regular amount of listener feedback using an optimized posting schedule.
3. The content makers make predictions about the post to determine whether a high arousal, i.e., a huge number of feedbacks, will exist. An ensemble-based classifier is recommended to infer the prediction rate of arousal that may yield to infer whether the new post has a high arousal or not.
4. Social media such as television, radio, and newspaper are used to analyze the user sentiment that explicates user feedback over different channel types.
5. A novel generation algorithm is proposed that uses association mining and NLP to compare the probabilistic topics. Besides, the proposed algorithm produces various subjective topics to achieve a higher precision ratio.

2. Literature survey

This section discusses the related works on information diffusion and information summarization of the social networks.

2.1 Information diffusion

Information diffusion has become one of the prevalent research topics in network analysis [13,14]. This work majorly addresses two key queries: (1) how does information diffuse in online social networks (OSN) [15,16]; and (2) how does information diffusion cause enhancement [16]. To overcome these two queries, information diffusion uses two standard models, namely, the independent cascade (IC) model [17,18] and the linear threshold (LT) model [19]. The former model has an opportunity to analyze a sender-centric/information push model to find the inactive neighbors. The latter model uses a receiver-centric/information pull model to examine the aggregate activated by the neighbors influencing the active user limits. For example, the spread of disease in a social network may be modeled using the IC model, whereas the spread of subjective opinion in social networks may be modeled using the LT model.

To disperse the information over a social network, data flow and dissemination of data is needed in the social network [20,21]. Topologic organization and time-based features are required to understand the nature of the social network [22]. In this chapter, three key factors such as network connection, signpost, and content poster are studied. Using a network connection, the user impact on a social network is determined to analyze the influential factors for WoM marketing. The preeminent posting time(s) is utilized to improve information diffusion. The arousing contents and the contents with factual sentiment are shown to upturn information diffusion.

2.1.1 User influences in OSN

In OSN, the stimulus of all users is not identical; it shows a discrepancy of connections and social location. Wu et al. [23] exhibited that > 1% of the social network creates 50% of social content, while the other users create inadequate content. Krishna et al. [24] identified that the influential users may influence the maximum level about a product or service. Thus, it is more indispensable to discover a few authorities that widely spread the information across the network. A maximum of powerful users or authorities may discover a social network to recognize the topologic structure [25]. Numerous researchers have studied the construction of social networks that solve the issue of stimulus extension [26]. The objective of the stimulus extension is to exploit product diffusion while decreasing the advancement rate through the selected subset of users called prominent users.

Further, numerous studies [27] attempted to explain the source of difficulties in finding the stimulus user using reputation trials such as Page-Rank [28], H-ITS [29], Z-score [30], Eigen-Vector [31] and Centrality-Measures [32], i.e., from link analysis. The existing works produce a static graph that is based on user connections to discover the maximum prominent nodes in the given graph. Conversely, they do not focus on the discovery of powerful influential users. As an authoritative situation of users varies through the sensitive topics [3], the prominent users are focused on topical content to create a dynamic graph based on social collaborations. In this chapter, topic-sensitive users are determined to advertise a popular subject topic through the WoM marketer.

As referred to by Ogilvy Cannes, 74% of users identify the WoM marketer to gain the decision factors. As referred to by MarketShare [33], WoM was shown to recover the effectiveness of the market, i.e., 54%. In this chapter, the topmost authorities or the WoM marketers are found to determine the network connection and topic sensitivities that propose a reinforcement concept to spread widely in the social network. The WoM marketer is more operative to improve user trustiness over the products if the multiple-WoM marketers mutually endorse the product within a community group. Comparable to the LT model, the marketers over multiple WoM marketers benefit to exceed the stimulus limit in a similar group.

2.1.2 Precise time to post

A post formed at a precise time may experience an upturn from the great feedback and thereby surge the dissemination. Behavioral user feedback alters through various social networks [34]. As an instance, on Twitter, the generation of user content is quite undersized compared to other social networks. Wu et al. [23] exhibited that all subjective content has a limited lifespan, which drops exponentially over a day. In this chapter, the users' feedback behavior represents a large social network, i.e., Facebook, to reveal the lifetime of post content that obtains the preponderance of feedback over a few hours of relocation. To learn users' behavioral feedback, connections and daily and weekly feedback forms are dynamically studied [35]. A few studies have been conducted for the factual analysis of the posting schedule.

The posting time uses dynamic social network users [36,37] to find the precise relocation, i.e., for individual users. It may use the posting schedules to derive social contacts and positions. Many features may affect listener feedback, such as features to analyze the topic contents [38] or features of the content creator [39].

2.1.3 Prediction popularity of content posts

Numerous works have studied the reputation of social media [40]. These works utilize the reputation of news posts to classify broadly into social connection–based and content-based. The content-based is further divided into post content and post sentiment. Social connection [41] uses societal features such as families and supporters to foresee content reputation. Zaman et al. [42] considered the feedback behavior of user retweets. They employed the author's information such as followers, source identities, etc. Suh et al. [43] analyzed various key factors to review the impact of retweets that showed the use of followers and friends to determine the impact of the tweets. Petrovic et al. [44] employed a passive-aggressive method to predict the retweet ratio that may lead to high information spread over sensitive users. Weng et al. [41] expected to analyze the imminent reputation of an article using prompt diffusion patterns. They decided that the community structure is based on article features to infer the maximum influential reputation.

2.1.3.1 Prediction popularity using content post

Several content-based methods are used to calculate the reputation of news posting content [45]. Bandari et al. [46] projected the reputation of content posts using prediction popularity. They measured four significant features, namely, the post category, subjective content of a new post, naming entities, and new source content. They applied classification and regression to predict the popularities. Lee et al. [47] intended to design a framework that predicts the comments over the observation of the articles. Tatar et al. [40] suggested a method to predict the rank of the new posts using users' comments. In contrast to the existence of the previous works, the social network involves different media channels to perform whether the channels have an arousal of news to publish or not. Arousal is very similar to the reputation of different user comments or feedback to derive the features from news articles. It relates to new coverage, reputation, and arousal prediction using embedded words.

2.1.3.2 Dynamic sentiment of new content posts

The sentiment of a new content post has become a key factor for the analysis of high information diffusion and reputation. Naveed et al. [8] exhibited 15 different sets of content features to predict the probability of a tweet. More attractive users may easily catch the WoM marketer's attention in applying logistic regression. Wu et al. [45] presented that the lifetime of undesirable posts is very limited over a longer time. They used classification techniques to predict the degeneration of social media content. However, in this chapter, it is shown that negative comments may lead to improve the user's reputation more than disinterested posts. Universidade Federal de Minas Gerais (UFMG) [48] developed an eminent tool to analyze the user's interest that relies on interest or polarization.

They graded each new article to popularize the sentiment score. Reis et al. [49] analyzed the article headlines to examine the popularity of a new communication channel. They presented the sentiment of article headlines to correlate with the reputation of the news report and undesirable comments. It may be posted autonomously to read the sentiment score of the headlines. On the other hand, they analyzed the sentiment polarity of a news post to associate with the reputation of content posts. The polarity of a news comment is not completely liberated by the polarity of the authentic post that functions as the divergence of the news post.

Zubiaga [50] discovered the difficulties in finding interesting topics in social media and a news channel such as The New York Times.

The author showed that the subjective topics mainly examine the unbreakable news such as political affairs and currency, whereas users are interested more in the posting of niche topics in social media. Common people may be interested to post niche news in social media that compares news/headlines to draw additional responsiveness.

2.2 Issue: information summarization

With the massive growth of generated posts, the content summarization is very difficult to summarize. It uses multiple methods, namely, snippets [51], hashtags [52], word cloud [53], and topics of interest [53], to review a large quantity of posted content. A quantity of documents summarize top representative sentences to discover the best concepts or interesting topics. The limited sentences may not offer hidden or latent topics that are beneficial to review the documents. It employs interesting topics to sense the diverse themes or key topics to understand and categorize the documents.

2.2.1 Interesting topics

Various statistical methods [54,55] have been suggested for the classification of interesting topics from the text documents. A topic model called Latent Dirichlet Allocation (LDA) [56] relies on the assumption of bag of words to analyze the impacts of machine learning and text mining. The LDA deliberates the document types such as the creation of topics and the multinomial word distribution to present the document types. Contrasted with probabilistic latent semantic analysis (PLSA) [57], the LDA model is a procreative model to create the subject topics. However, the LDA may not create expressive phrases to represent many universal words that may not deliver comprehensive information. As an instance, the LDA makes several inconsequential words such as "framework," "information," "method," "model," and "type" using a probability of eminent topics from the machine learning.

To express textual documents, numerous phrases have been recommended [58]. Phrase-Discovering Latent Dirichlet Allocation (PD-LDA) and Topical N-Gram are chosen to discover the phrases. These phrasing methods discover meaningful phrases but experience more complexity and inadequate scalability. Besides, the generated topics include various existing topics to categorize the modeling methods but are not perfect to display the generated topics. They are not comprehensible but often redundant. In the experiment, these phrasing methods generate fewer interpretable phrases such as "experiential study," "constructed malware," and "incidental study" to label the interesting topics such as research discussions and research extents.

3. Research guidelines

Adequate research is practiced to recognize the processes of information diffusion. The detailed investigation concludes the uncomplicated information diffusion to classify the issues such as the 3 Ws i.e., what, why, and where. At first, the term "*what*" refers to the query that determines the latent information instituted in social networks. As an instance, a huge customer data contains some stimulating discoveries. The second term of "*why*" represents the query used to determine the use of propagated information to visualize the data connections or information. The third term of "*where*" represents the diffused information to classify the influential users. Assume that A posts the information, both B and C may diverse the collective information, which are influenced to propagate the information on the network.

These features support the node to understand the destination, i.e., for future diffusion and to predict future information diffusion. "*what*" and "*why*" include the features of information diffusion to relate the prediction with the "*where*" query. In reality, the "*3 Ws issue*" signifies the use of information diffusion in research. Fig. 5.2 shows guidelines of information diffusion that use data extraction and storage to describe the process of diffusion stages and to examine the stimulus factors regulating the information diffusion.

Based on the research guidelines, the collected works related to these subjects can be categorized into explanatory and predictive models. The most extensive study uses basics and their accessible model is divided into studied, investigated, discussed, and linked. Lastly, imminent challenges and approaches present the research in the form of information diffusion. These models are based on the view of the independent literature review to complement the nature of the models. Fig. 5.3 shows the classification of the information diffusion model.

3.1 Explanatory model

Basic information is widely dispersed to analyze the interactions between dissimilar individuals in societies. These characters can be observed as the social nodes that abstract the representation of real-time entities. The mutual interaction between the entities defines relations among the network edges to run two cooperative nodes

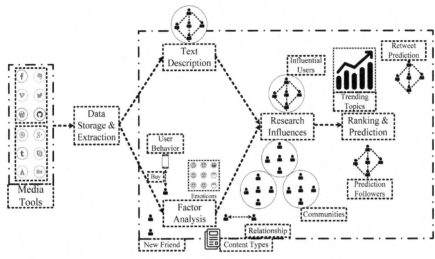

FIGURE 5.2

Research guidelines of information diffusion.

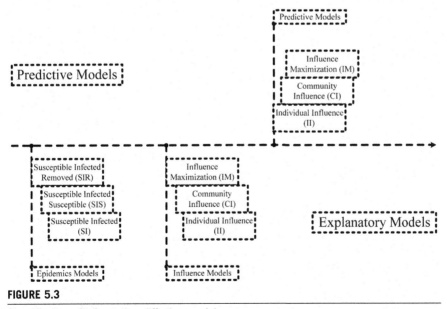

FIGURE 5.3

Classification of information diffusion model.

in social networks. Thus, a social group may mutually be represented to disseminate a huge social network into a piece of information. The dissimilar groups of communities have a diverse characteristic to define the network as homogenous and heterogeneous. This model aims to inspect the process of information diffusion to elucidate the stimulus factors that affect the phenomenal attempts.

3.2 Epidemics model

The process of information diffusion deliberates the epidemic spread. In the transmission of epidemics, the infected users with pathogens are vulnerable to the production agent. The infection may divide the septic users from vulnerable users that diffuse the information using communicators that investigate the recipients' information. It uses the diffused information to learn the basic epidemic models. The compartment model of epidemics classifies into three basic models, namely, susceptible-infected (SI), susceptible-infected-susceptible (SIS), and susceptible-infected-recovered (SIR).

3.2.1 SI model

The influential rate of birth and death does not rely on the available number of individuals. This model adopts that the overall number of individuals is N. It is basically distributed into two specific labels: S for susceptible and I for infected. At time t, $s(t)$ signifies the vulnerable quantity of the entire people, $i(t)$ signifies the infested quantity, and λ signifies the regular connection rate, which represents the quantity of the vulnerable users over infected users in the overall population, where $s(t) + i(t) = 1$. Hence, $N_S(t) + N_i(t) = N$.

3.2.2 SIS model

The SI model is unrealistic, as it is not allowed to cure the infected users after being infected. Thus, the SIS model reports to solve the issue of intricate and adaptive networks. The infected parameters such as N, $s(t)$, $i(t)$, and λ are additionally assumed to represent the daily rates, i.e., μ. In other words, μ defines the quantity of the infected users cured over the available population. The incremental change can be expressed as follows: $N \cdot \frac{di}{dt} = \lambda \cdot N_{s_i} - \mu \cdot N_i$, where $\lambda \cdot N_{s_i}$ is the incremental number of patients each day, and $\mu \cdot N_i$ is the incremental number of treated patients each day.

3.2.3 SIR model

The SIR model joined with differential self-motivated equivalences to establish whether an individual is cured over immunity or not. This is not reserved to divide the overall population N into S, I and R to represent the vulnerable and infested users to describe in two models where R signifies invulnerable users, and $s(t) + i(t) + r(t) = 1$. It is assumed that $s\langle 0 \rangle = s_0, i_0$, $r\langle 0 \rangle = 0$, and $\frac{ds}{dt} + \frac{di}{dt} + \frac{dr}{dt} = 0$. The regular growth of the cumulative amount of invulnerable users is stated as $N \cdot \frac{dr}{dt} = \mu \cdot N_i$.

The assessment of basic epidemic models is presented in Fig. 5.4. It exhibits the distribution procedure of an illness in an endemic that shows the user prominence in social networks.

3.3 Influence models

The influence analysis classifies three significant factors such as individual, community, and maximization.

3.3.1 Individual influence

Individual influence is the opinion related to the leader-related research that shows as a bridge of information diffusion. It has a certain influence over other users that cannot be ignored in the research of information diffusion. The opinion leaders include a network structure, mutual information, and user attributes to measure the importance of the community nodes. Page rank and other procedures are used to rank the listener nodes. This technique is very guileless but the accuracy is not high. The mutual information exchange is considered to focus on the community groups.

Table 5.1 summarizes the comparisons of the individual influence methods. It is divided into three aspects: network construction, user collaborations, and user characteristics. The individual influence has one element to include a tipping-point node [59]. It may include several key elements, such as user action and centrality [60], movement and contact time dissemination [61], and movement and associations [62,63]. It is very significant to analyze the influential research that combines three elements to produce more specific information [62]. Unrelatedly, they have a measurable criterion to weight the influence factor. The condition contains out-degree [59], activists [60,63], centrality and intermediary [39], diffusion capability [61], and core coverage ratio [62]. These methods always have a rough selection-

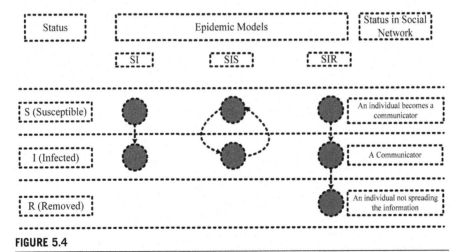

FIGURE 5.4

Assessments of three basic epidemic models.

Table 5.1 Comparisons of individual influence methods.

Existing work	Structured network	Attributes			Method used	Criterion	Preferred application
		Interaction	Behavior	Features			
Wang et al. [59]	Yes	–	–	–	Social network analysis	Out-degree	Identify the opinion leaders to analyze the prediction
Chen et al. [60]	–	Yes	Yes	Centrality	Competency level	Centrality, activists	Identify the opinion leaders to analyze the influence maximization
Mao et al. [61]	Yes	–	Yes	Access time	Social network analysis	Diffusion capability	Influence the prediction
Wu et al. [62]	Yes	Yes	Yes	Weight and topics	Page rank	Core coverage ratio	Mining on the opinion leader
Ullah [63]	Yes	Yes	Yes	Neighbor to neighbor	Social network analysis	Activists	Identify the influential nodes

based network structure for opinion leaders first. The user interactions and key attributes are broken down to compute more precise results. In the past, the interaction-based individual influences [61−63] has attracted researcher attention.

3.3.2 Community influence

A community has a group of folks with more or less collective properties to form various communities on the basics of interested groups. A community has a network subset to connect the users densely with similar attributes, e.g., like, comment, and share. The construction of social networks may periodically change to form relatively stable communities. The main task is to sense the communities that might have a high influential factor within a network. Several methods have been suggested to analyze the links and attributes of the networks.

Yang et al. [64] suggested a technique based on a discriminative probabilistic model. This technique uses PCL-DC to analyze the group communities that associate with links and content to describe the content popularity. Zhou et al. [65] suggested a virtual property based on the node insertion that uses the SA-Cluster-Inc method to calculate the community distance between the nodes. It has a new attribute graph to discover the node edges. A K-means cluster is highly preferred to gather the new nodes. Ruan et al. [66] also combined the viewpoint to link the contents, but it is not as effectual as it might be. An efficient COD-ICIL method was proposed to detect the communities that combine links and contents to strengthen the community relationship.

Yang et al. [67] shared a significant perspective to choose the naming attributes that propose a community with edge structure and node attributes. Yang and Manandhar [68] considered links of either content or attributes to determine the sentiment topic that forms the communities to associate contacts, subjects, and emotions. These are employed to form different communities with the distributed topics. Peng et al. [69] deliberated to connect nearly 10 connections that associate the relative nodes to form a community. Though it has less connectivity in the initial graph, it may represent a better graph structure. Gurini et al. [70] decided to use a sentiment-based technique called SVO method that considers the target attitudes of the user to solve the objective-related contents. This method uses $\theta = 0.8$ to represent a connection threshold.

Ullah et al. [71] planned a suitable model to sense the group communities that compute the interest similarity. The social network measures the trust between the source node and the target node. Importantly, the value of trust is bidirectional. Therefore, the researchers practice a cluster-based method that predominantly builds a network structure based on attributes or contents. The initial structure updates the links to increase the precision factor of community detection and moderate the consumption of time. From the literature, it is observed that the model performance may not be upgraded after the consideration of reasonable attributes. As an instance, a sentiment analysis cannot be fit to detect whether the communities can be useful to examine the situation or not. Table 5.2 summarizes the comparison of community detection.

Table 5.2 Comparison of community detection.

Existing model	Network link	Contents or attributes	Sentiment	Technique used	Measurable standard
Yang et al. [64]	Yes	Yes	–	Probability	–
Zou et al. [65]	Yes	Topic and prolific	–	Cluster	Entropy and density function
Ruan et al. [66]	Yes	Tag, context, title, and word	–	Cluster	Quality function
Yang et al. [67]	Yes	Text, user	Yes	Probability	Topic similarity
Gurini et al. [70]	Yes	Interest	Yes	Cluster	Homophily
Ullah et al. [71]	Yes	Trust	–	Both probability and cluster	Quality function

3.3.3 Influence maximization

The influence maximization includes two phases: selection (training) and action (competition) to select a suitable round, i.e., single or multiple. The multiple round uses the historical influence to appraise the seed nodes, whereas the individual influence maximizes the use of certain topics or a piece of content information. However, it is referred to as competitive research to highlight the use of multiple items or information.

The central objective is to exploit the group influence to find whether the adversary's approach is known or unknown. The literature analysis reveals that the influence maximization primarily focuses on both the levels, i.e., individual and community. These levels use the shared objects to discover the seed nodes, whereby their influence may be maximized. Table 5.3 shows a comparison of influence maximization methods.

4. Predictive model

This model uses futuristic information to predict the diffusion process that has certain key factors in social networks. The models are often employed to maximize the influential features of three models, namely, IC, LT, and game theory (GT).

4.1 Model 1: independent cascade

In the IC model, the inactive node v can be activated by the active node u independently with a probability of P_U, v over a time t. If a node v is initiated, then v will be

Table 5.3 Comparison of influence maximization methods.

Existing model	Seed detection	Technique used	Data/ model	Multiple round	Innovation/ information
Lei et al. [72]	Yes	Explore, exploit, and heuristic	Model	Yes	—
Lin et al. [73]	Yes	Tool maximization	Data	Yes	Yes
Horel et al. [74]	Yes	Paradox	Data	—	—

an active node over time t. Unrelatedly, u activates v over time t, but u will not activate v over the execution time. This model mainly applies prediction and influence research to determine influential dispersion.

4.2 Model 2: linear threshold

In the LT model, v is an active node to set the activation threshold over time t. v tries to activate v to find the degree of influence, where the active nodes surpasses the threshold limit v and the inactive node v finds an active node over time $(t + 1)$. The active neighbors may stimulate v times to study the degree of influence in social networks. The threshold behavior is chosen during the dispersion process to find the collective effect of the influence factors.

4.3 Model 3: game theory

In the GT model, it has a strategy of maximum profit to limit multiple or groups of individuals with detailed constraints. It continually utilizes an adversary strategy to exploit the profits. A piece of material is chosen to infer the effect of overheads, profits, and intentional choice.

The network structures and the consumer behaviors are preferred independently to combine the models. A social network is a dynamic network that uses a robust prediction to apply in a social network. The literature analysis observes the comparative features of three models, such as IC, LT, and GT. The IC is generally sender-centric that considers the sender information, whereas LT is receiver-centric, and GT is more neutral-centric to consider the network profit. They will endeavor to do a suitable study to analyze the nature of dynamic networks. Table 5.4 shows the comparison of predictive models.

5. Future directions

Information diffusion is becoming an attractive topic in social networks that have innovative studies to address the resolved issues. Fig. 5.5 shows the future challenges, issues, and methods in information diffusion.

Table 5.4 Comparison of predictive models.

Existing model	Predictive model			Researcher knowledge	Application domain
	IC	**LT**	**GT**		
Arora et al. [75]	Yes	–	–	Likelihood information using diffusion episodes	Prediction probability
Barbieri et al. [76]	Yes	Yes	–	Topic aware	Topic prediction
Chen et al. [77]	–	Yes	–	Influential nodes and activation threshold	Select the influential nodes
Wang et al. [78]	–	–	Yes	Information behavior at a micro level	Predicting the relationship

•   Importance of weak nodes

This research finds the seed nodes to analyze the factors including degree, closeness, and betweenness of the centrality nodes. These metrics are usually explicit. The implicit weak nodes do not influence the degree of centrality nodes to play an essential part in information diffusion over a social network.

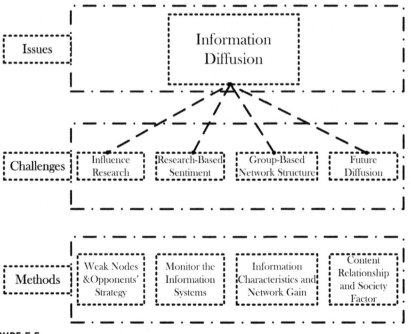

FIGURE 5.5

Future challenges, issues, and methods in information diffusion.

- A competitive maximization in influence diffusion

The recent research often focuses on a single piece of information that has various kinds of information containing advertised information to identify the interest of stakeholders. As a competitive maximization, the information is contained in certain advertisements referred to as a modest stimulus. The competitive influence may be applied in the competitive community to infer positive or negative statements in the social network.

- A group status with network structure

This model aims to analyze the group status that greatly affects the surrounding nodes to identify the group characteristics in the information diffusion. So, this model is more impartial to examine the several key factors to consider the network structure. It mainly depends upon information diffusion to inspect whether there is a threshold gain to meet the objective of the social network.

- Predicting the information diffusion

The predictions symbolize the use of OSN to infer the impacts of media tools such as television, newspapers, and other traditional media. To practice the prediction research, a long-term process is preferred that shows the future directions.

6. Discussion and conclusion

In the past, various social media tools have emerged in contact with individual perspectives conveniently. An enormous information quantity may therefore be formed to solve the major issues that prefer multidisciplinary research such as computer science, sociology, psychology, and economics. Researchers build confident models to describe the diffusion phenomenon that predict the future diffusion using machine learning. In this chapter, the predictive and explanatory models were detailedly investigated for information diffusion. The main objective is to demonstrate the processes of information diffusion that has epidemic models such as SI, SIS, and SIR to simulate data diffusion. Therefore, additional prominences are contributed to examine the complex structure, shared information, and consumer behaviors.

With the progress of research, the imminent transmission process may influence three prediction models, namely, IC, LT, and GT, to examine the research applications. These models may use the predictive models based on basic instructive models to reveal the nature of the dynamic graph. It employs the propagation process to reveals the effect of opinion leaders. This model is then used to build the network structure based on the given social network analysis. A logical analysis involves information diffusion to divide the models into explanatory and predictive. They are mostly used to analyze three points: (1) aim of information diffusion models; (2) finding the factors or predicting the future outcomes or direction; and (3) models not independent to explore the existing issues.

References

[1] D. Easley, J. Kleinberg, Networks, Crowds, and Markets: Reasoning about a Highly Connected World, Cambridge University Press, 2010.

[2] C.-T. Li, Y.-J. Lin, M.-Y. Yeh, Forecasting participants of information diffusion on social networks with its applications, Inf. Sci. 422 (2018) 432–446.

[3] A. Wagener, Hypernarrativity, storytelling, and the relativity of truth: digital semiotics of communication and interaction, Postdigit. Sci. & Edu. 2 (1) (2020) 147–169.

[4] A.K. Kushwaha, A.K. Kar, P.V. Ilavarasan, Predicting information diffusion on Twitter a deep learning neural network model using custom weighted word features, in: Responsible Design, Implementation and use of Information and Communication Technology, vol. 12066, 2020, p. 456.

[5] R. Nugroho, C. Paris, S. Nepal, J. Yang, W. Zhao, A survey of recent methods on deriving topics from Twitter: algorithm to evaluation, Knowl. Inf. Syst. (2020) 1–35.

[6] A. Orben, Teenagers, screens and social media: a narrative review of reviews and key studies, Soc. Psychiatr. Psychiatr. Epidemiol. (2020) 1–8.

[7] N. Kumar, G. Ande, J.S. Kumar, M. Singh, Toward maximizing the visibility of content in social media brand pages: a temporal analysis, Soc. Netw. Anal. & Min. 8 (1) (2018) 11.

[8] N. Naveed, T. Gottron, J. Kunegis, A.C. Alhadi, Bad news travel fast: a content-based analysis of interestingness on Twitter, in: Proceedings of the 3rd International Web Science Conference, ACM, 2011, p. 8.

[9] N. Kumar, A. Yadandla, K. Suryamukhi, N. Ranabothu, S. Boya, M. Singh, Arousal prediction of news articles in social media, in: International Conference on Mining Intelligence and Knowledge Exploration, Springer, 2017, pp. 308–319.

[10] N. Kumar, R. Nagalla, T. Marwah, M. Singh, Sentiment dynamics in social media news channels, Online Soc. Netw. & Media 8 (2018) 42–54.

[11] J. Prüfer, P. Prüfer, Data science for entrepreneurship research: studying demand dynamics for entrepreneurial skills in The Netherlands, Center Discussion Paper, Small Bus. Econ. (2019) 36, 2019–005.

[12] N. Kumar, R. Utkoor, B.K. Appareddy, M. Singh, Generating topics of interests for research communities, in: International Conference on Advanced Data Mining and Applications, Springer, 2017, pp. 488–501.

[13] N. Chouchani, M. Abed, Online social network analysis: detection of communities of interest, J. Intell. Inf. Syst. 54 (1) (2020) 5–21.

[14] M. Schoenfeld, J. Pfeffer, Networks and context: algorithmic challenges for context-aware social network research, in: Challenges in Social Network Research, Springer, Cham, 2020, pp. 115–130.

[15] K. Shu, H.R. Bernard, H. Liu, Studying fake news via network analysis: detection and mitigation, in: Emerging Research Challenges and Opportunities in Computational Social Network Analysis and Mining, Springer, 2019, pp. 43–65.

[16] V. Arnaboldi, M. Conti, A. Passarella, R.I. Dunbar, Online social networks and information diffusion: the role of ego networks, Online Soc. Netw. & Media 1 (2017) 44–55.

[17] L. Yang, Z. Li, A. Giua, Containment of rumor spread in complex social networks, Inf. Sci. 506 (2020) 113–130.

[18] X. Wang, K. Deng, J. Li, J.X. Yu, C.S. Jensen, X. Yang, Efficient Targeted Influence Minimization in Big Social Networks, World Wide Web, 2020, pp. 1–18.

[19] M. Alshahrani, Z. Fuxi, A. Sameh, S. Mekouar, S. Huang, Efficient algorithms based on centrality measures for identification of Top-K influential users in social networks, Inf. Sci. 527 (2020) 88−107.

[20] L. Liu, X. Wang, Y. Zheng, W. Fang, S. Tang, Z. Zheng, Homogeneity trend on social networks changes evolutionary advantage in competitive information diffusion, New J. Phys. 22 (1) (2020) 013019.

[21] R. Yan, Y. Zhu, D. Li, Y. Wang, Community based acceptance probability maximization for target users on social networks: algorithms and analysis, Theor. Comput. Sci. 803 (2020) 116−129.

[22] L. Jain, R. Katarya, S. Sachdeva, Opinion leader detection using whale optimization algorithm in online social network, Expert Syst. Appl. 142 (2020) 113016.

[23] S. Wu, J.M. Hofman, W.A. Mason, D.J. Watts, Who says what to whom on Twitter, in: WWW, ACM, 2011, pp. 705−714.

[24] R. Krishna, C.M. Prashanth, Identifying influential users on social network: an insight, in: Data Management, Analytics and Innovation, Springer, Singapore, 2020, pp. 489−502.

[25] A. Raychaudhuri, S. Mallick, A. Sircar, S. Singh, Identifying influential nodes based on network topology: a comparative study, in: Information, Photonics and Communication, Springer, Singapore, 2020, pp. 65−76.

[26] S.M.C. Loureiro, R.G. Bilro, Be or not Be online engaged: exploring the flow from stimuli to e-WOM on online retail consumers, in: Exploring the Power of Electronic Word-of-Mouth in the Services Industry, IGI Global, 2020, pp. 18−34.

[27] S.H.W. Chuah, D. El-Manstrly, M.L. Tseng, T. Ramayah, Sustaining customer engagement behavior through corporate social responsibility: the roles of environmental concern and green trust, J. Clean. Prod. (2020) 121348.

[28] G. Di Tommaso, M. Gatti, M. Iannotta, A. Mehra, G. Stilo, P. Velardi, Gender, rank, and social networks on an enterprise social media platform, Soc. Network. 62 (2020) 58−67.

[29] J.A. Lee, M.S. Eastin, I like what she's# endorsing: the impact of female social media influencer's perceived sincerity, consumer envy, and product type, J. Interact. Advert. (2020) 1−39 (just-accepted).

[30] T. Üsküplü, F. Terzi, H. Kartal, Discovering activity patterns in the city by social media network data: a case study of istanbul, Appl. Spat. Anal. & Policy (2020) 1−14.

[31] A. Muruganantham, G.M. Gandhi, Framework for social media analytics based on multi-criteria decision making (MCDM) model, Multimed. Tool. Appl. 79 (5) (2020) 3913−3927.

[32] K. Chakma, R. Chakraborty, S.K. Singh, Finding correlation between Twitter influence metrics and centrality measures for detection of influential users, in: Computational Intelligence in Data Mining, Springer, Singapore, 2020, pp. 51−62.

[33] G. Paré, J. Marsan, M. Jaana, H. Tamim, R. Lukyanenko, IT vendors' legitimation strategies and market share: the case of EMR systems, Inf. & Manag. (2020) 103291.

[34] M.J. Lubbers, A.M. Verdery, J.L. Molina, Social networks and transnational social fields: a review of quantitative and mixed-methods approaches, Int. Migrat. Rev. 54 (1) (2020) 177−204.

[35] A.M. Thompson, W. Wiedermann, K.C. Herman, W.M. Reinke, Effect of daily teacher feedback on subsequent motivation and mental health outcomes in fifth grade students: a person-centered analysis, Prev. Sci. (2020) 1−11.

[36] N. Spasojevic, Z. Li, A. Rao, P. Bhattacharyya, When-to-post on social networks, in: SIGKDD, ACM, 2015, pp. 2127–2136.

[37] A. Zarezade, U. Upadhyay, H.R. Rabiee, M. Gomez-Rodriguez, Redqueen: an online algorithm for smart broadcasting in social networks, in: WSDM, ACM, 2017, pp. 51–60.

[38] J.C.S. dos Rieis, F.B. de Souza, P.O.S.V. de Melo, R.O. Prates, H. Kwak, J. An, Breaking the news: first impressions matter on online news, in: ICWSM, 2015, pp. 357–366.

[39] R. Yan, C. Huang, J. Tang, Y. Zhang, X. Li, To better stand on the shoulder of giants, in: Proceedings of the 12th ACM/IEEE-CS Joint Conference on Digital Libraries, ACM, 2012, pp. 51–60.

[40] A. Tatar, P. Antoniadis, M.D. De Amorim, S. Fdida, From popularity prediction to ranking online news, Soc. Netw. Anal. & Min. 4 (1) (2014) 174.

[41] L. Weng, F. Menczer, Y.-Y. Ahn, Predicting successful memes using network and community structure, in: ICWSM vol. 8, 2014, pp. 535–544.

[42] T.R. Zaman, R. Herbrich, J. Van Gael, D. Stern, Predicting information spreading in Twitter, in: Workshop on Computational Social Science and the Wisdom of Crowds, NIPS. Citeseer, 2010, pp. 17 599–601.

[43] B. Suh, L. Hong, P. Pirolli, E.H. Chi, Want to be retweeted? large scale analytics on factors impacting retweet in Twitter network, in: IEEE Second International Conference on Social Computing, IEEE, 2010, pp. 177–184.

[44] S. Petrovic, M. Osborne, V. Lavrenko, Rt to win! predicting message propagation in Twitter, in: ICWSM, vol. 11, 2011, pp. 586–589.

[45] S. Wu, C. Tan, J.M. Kleinberg, M.W. Macy, Does bad news go away faster?, in: ICWSM, 2011, pp. 646–649.

[46] R. Bandari, S. Asur, B.A. Huberman, The pulse of news in social media: forecasting popularity, in: ICWSM, 2012, pp. 26–33.

[47] J.G. Lee, S. Moon, K. Salamatian, An approach to model and predict the popularity of online contents with explanatory factors, in: Web Intelligence and Intelligent Agent Technology (WI-IAT), 2010 IEEE/WIC/ACM International Conference on, vol. 1, IEEE, 2010, pp. 623–630.

[48] T. Silva, A. Loureiro, J. Almeida, P. de Melo, Large scale study of city dynamics and urban social behavior using participatory sensor networks, in: Anais do XXVIII Concurso de Teses e Dissertações, SBC, February 2020, pp. 13–18.

[49] J. Reis, P. Gonçalves, P. Vaz de Melo, R. Prates, F. Benevenuto, Magnet news: you choose the polarity of what you read, in: ICWSM, 2014, pp. 652–653.

[50] A. Zubiaga, Newspaper editors vs the crowd: on the appropriateness of front page news selection, in: WWW, ACM, 2013, pp. 879–880.

[51] C. Li, Y. Duan, H. Wang, Z. Zhang, A. Sun, Z. Ma, Enhancing topic modeling for short texts with auxiliary word embeddings, ACM Trans. Inf. Syst. 36 (2) (2017) 11.

[52] W. Cui, Y. Wu, S. Liu, F. Wei, M.X. Zhou, H. Qu, Context preserving dynamic word cloud visualization, in: Visualization Symposium (PacificVis), 2010 IEEE Pacific, IEEE, 2010, pp. 121–128.

[53] X. Wang, A. McCallum, X. Wei, Topical n-grams: phrase and topic discovery, with an application to information retrieval, in: ICDM, IEEE, 2007, pp. 697–702.

[54] Y. Hu, Z. Zhou, K. Hu, H. Li, Detecting overlapping communities from micro blog network by additive spectral decomposition, J. Intell. Fuzzy Syst. 38 (1) (2020) 409–416.

[55] S.R. Sahoo, B.B. Gupta, Fake profile detection in multimedia big data on online social networks, Int. J. Inf. Comput. Secur. 12 (2−3) (2020) 303−331.

[56] I. Sutherland, Y. Sim, S.K. Lee, J. Byun, K. Kiatkawsin, Topic modeling of online accommodation reviews via latent dirichlet allocation, Sustainability 12 (5) (2020) 1821.

[57] Z. Zhou, X. Zhang, Z. Guo, Y. Liu, Visual abstraction and exploration of large-scale geographical social media data, Neurocomputing 376 (2020) 244−255.

[58] L. Lin, Y. Rao, H. Xie, R.Y.K. Lau, J. Yin, F.L. Wang, Q. Li, Copula guided parallel gibbs sampling for nonparametric and coherent topic discovery, IEEE Trans. Knowl. Data Eng. (2020) 1−16.

[59] C.X. Wang, X.H. Guan, T. Qin, Y.D. Zhou, Modelling on opinion leader's influence in microblog message propagation and its application, J. Softw. 26 (2015) 1473−1485.

[60] B. Chen, X. Tang, L. Yu, Y. Liu, Identifying method for opinion leaders in social network based on competency model, J. Commun. 35 (2014) 12−22.

[61] J.X. Mao, Y.Q. Liu, M. Zhang, S.P. Ma, Social influence analysis for micro-blog user based on user behavior, Chin. J. Comput. 37 (2014) 791−800.

[62] X. Wu, H. Zhang, X. Zhao, B. Li, C. Yang, Mining algorithm of microblogging opinion leaders based on user-behavior network, Appl. Res. Comput. 32 (2015) 2678−2683.

[63] F. Ullah, S. Lee, Identification of influential nodes based on temporal-aware modeling of multi-hop neighbor interactions for influence spread maximization, Physica A 486 (2017) 968−985.

[64] T. Yang, R. Jin, Y. Chi, S. Zhu, Combining Link and Content for Community Detection, Springer, New York, NY, USA, 2014, pp. 190−201.

[65] Y. Zhou, H. Cheng, J.X. Yu, Clustering large attributed graphs: an efficient incremental approach, in: Proceedings of the 2010 IEEE International Conference on Data Mining, Sydney, Australia, 13−17 December 2010, pp. 689−698.

[66] Y. Ruan, D. Fuhry, S. Parthasarathy, Efficient community detection in large networks using content and links, in: Proceedings of the 22nd International Conference on World Wide Web, Rio de Janeiro, Brazil, 13−17 May 2012, pp. 1089−1098.

[67] J. Yang, J. Mcauley, J. Leskovec, Community detection in networks with node attributes, in: Proceedings of the 2013 IEEE 13th International Conference on Data Mining Workshops, Dallas, TX, USA, 7−10 December 2013, pp. 1151−1156.

[68] B. Yang, S. Manandhar, Community discovery using social links and author-based sentiment topics, in: Proceedings of the 2014 IEEE/ACM International Conference on Advances in Social Networks Analysis and Mining, Beijing, China, 17−20 August 2014, pp. 580−587.

[69] C. Peng, T.G. Kolda, A. Pinar, Accelerating Community Detection by Using K-Core Subgraphs, arXiv, 2014 arXiv:11403.2226.

[70] D.F. Gurini, F. Gasparetti, A. Micarelli, G. Sansonetti, Analysis of sentiment communities in online networks, in: Proceedings of the International Workshop on Social Personalisation & Search Co-located with the ACM SIGIR Conference, Santiago, Chile, 9−13 August 2015, pp. 1−3.

[71] F. Ullah, S. Lee, Community clustering based on trust modeling weighted by user interests in online social networks, Chaos Solit. Fractals 103 (2017) 194−204.

[72] S. Lei, S. Maniu, L. Mo, R. Cheng, P. Senellart, Online influence maximization, in: Proceedings of the 21th ACM SIGKDD International Conference on Knowledge Discovery and Data Mining, Sydney, NSW, Australia, 10−13 August 2015, pp. 645−654.

[73] S.C. Lin, S.D. Lin, M.S. Chen, A learning-based framework to handle multi-round multi-party influence maximization on social networks, in: Proceedings of the 21st

ACM SIGKDD Conference on Knowledge Discovery and Data Mining, Sydney, NSW, Australia, 10–13 August 2015, pp. 695–704.

[74] T. Horel, Y. Singer, Scalable methods for adaptively seeding a social network, in: Proceedings of the 24th International World Wide Web Conference (WWW2015), Florence, Italy, 18–22 May 2015, pp. 1–14.

[75] A. Arora, S. Galhotra, S. Virinchi, S.A. Roy, A scalable algorithm for influence maximization under the independent cascade model, in: Proceedings of the 24th ACM International Conference on World Wide Web Companion, Florence, Italy, 18–22 May 2015, pp. 35–36.

[76] N. Barbieri, F. Bonchi, G. Manco, Topic-aware social influence propagation models, Knowl. Inf. Syst. 37 (2012) 555–584.

[77] H. Chen, Y.T. Wang, Threshold-based heuristic algorithm for influence maximization, J. Comput. Res. Dev. 49 (2012) 2181–2188.

[78] Y. Wang, J. Yu, W. Qu, H. Shen, X. Cheng, C. Lin, Everlutionary game model and analysis methods network group behavior, Chin. J. Comput. 38 (2015) 282–300.

Forensic analysis in social networking applications

6

B.D. Deebak[1], Hadi Zahmatkesh[2]

[1]*School of Computer Science and Engineering, Vellore Institute of Technology, Vellore, Tamil Nadu, India;* [2]*Department of Mechanical, Electronic and Chemical Engineering, OsloMet - Oslo Metropolitan University, Oslo, Norway*

1. Introduction

In the investigation of a criminal offense, social network content commonly analyzes domestic and monetary disputes. To regularize the incidental process, legal practitioners and law enforcement agencies are involved. The relevant information of the individual is collected to analyze the suspicious evidence precisely. It has an interesting assignment to process the complexity of forensic records. Moreover, it is massive in volume and dynamic in network structure to process the technical data. Notably, this practice is rightfully exciting the discretion laws and joint ownership. The digital device syndicates extensive practice and operational activities proliferate the use of massive electronic data. An individual may create a profile of social media on digital devices to analyze and investigate the electronic evidence.

Without maintaining the automated tools, the existing methods address several challenges including inefficient task management, gathering technical evidence, and constituting the legal rights. The forensic evidence is permitted to exercise several approaches that collect crucial requirements such as integrity, reliability, and duplicability to design an automation tool [1]. The legal proceedings criticize the systematic fundamentals and rigorous testing [2] to emphasize the use of proven strategies as a substitute for deep-rooted analytical techniques. Data mining approaches use automated techniques that rely on natural language processing. However, they are not intended for digital forensics because of the legal limitations of data provenience [3]. These technical approaches are constructed using statistics and estimation, so they cannot make evident the origin of data.

To investigate the criminal evidence, social media content gives particular support to explore the potential rights of suspects, victims, and witnesses. The social contents including posted contents, friend lists, images, geolocation data, and videos dealing with dynamic data subsets to create metadata that holds the criminal investigation. Additionally, it may be used to authenticate the obtained evidence that has a legal requirement to inspect the digital devices. The device data traces the criminal histories to perform network analysis, performed by the digital examiners. The current practices include the investigation of social profiles to simplify information access.

However, the investigative analysis creates an additional workload to analyze the source of social networks.

Therefore, digital analysts cannot deal with massive data and extensive interaction to proliferate the individual activities. Various social network platforms experience inherent irregularities and difficulties to investigate the digital data more efficiently. They usually conclude with incomplete and independent pieces of evidence that deliver imperfect information in connection with real-time events. A promising set of data should be handled meticulously to investigate the scientific issues. Thus, this chapter reviews the current state of research to outline legal and practical difficulties that majorly include the extraction and protection of social evidence. The critical review shows the strength of existing solutions to identify the primary objectives, current challenges, and future directions in social media and digital forensics. Additionally, this work classifies the openings in current methodologies to highlight the critical investigation to deal with the open issues addressed in the network domains.

1.1 Unparallel opportunities

Regardless of internet benefits, the emerging evolution has made incomparable openings for criminality and mismanagement. The connected environments have some properties to simplify and reassure illegal behavior. The geographic locations are efficiently connected to form the communities that relate the individual relationships, follower interest, and essential media sources [4,5]. Also, user anonymity may be useful to ease contact risk, therefore reassuring illegal activities [6]. The individual opportunities play a significant characteristic of inhibition loss to influence the personality behavior to keep real-life identity discrete. With the strength of deviant characters i.e., fraudsters, cyberstalkers, the individual may use social networks or online media tools to prepare the possible fatalities or to observe any specific behavior of the targets.

Their illegal objectives may be unknown to provoke the naive users into a deceptive act [7]. Correspondingly, the criminals may infiltrate online profiles to obtain the benefit of monetary gain that conforms with diverse scamming techniques to stimulate financial payments, services, or investments [8].

1.2 Nature of cyber crime

The online environment reassures the individuals to contribute to the source of danger. The online communities share their personal information such as interest, gender, and location on social media, where millions of abusers may learn these private pieces of information to carry out malicious activities [9]. The online strangers interact with private matter and personal information to cause some potential risks to the target users [10]. The other criminal behavior arouses severe discrimination to recognize the use of social websites, online disclosures, shopping, and file maltreatment [11]. The digital or computer-based crime is called cybercrime

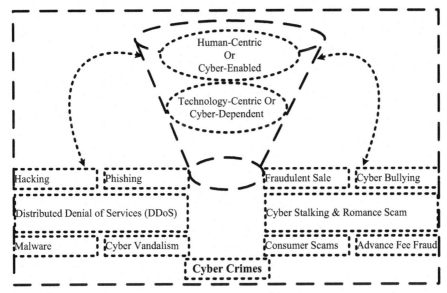

FIGURE 6.1

Classification of cyber crimes.

and refers to the criminal behavior through internet facilities and digital devices [12]. Wall presented the organization of the crime to recognize criminal behaviors including cyber violence, cyber trespassing, cyber obscenity, and cyber theft [13].

Cyber violence includes online behavior as a part of sensitive damages and physical, and emotional issues to the individual or community groups. Cyber trespassing is associated with several offensive techniques to gain unauthorized system access. Cyber obscenity considers the deviant or nondeviant behavior in the online environment. Cyber theft uses utilization techniques for illegal acquisition, copyrighted materials, and media frauds (Fig. 6.1).

1.3 Addressing issues

Technologic reliance increases the use of the internet in daily life, which leads to an obvious rise in e-crimes. Cifas showed that UK fraud prevention services examined 324,683 QUOTE QUOTE fraudulent cases in 2016, which rose from 1.2%in 2015. This service found that 53% were connected with personality fraud. Moreover, 88% were simplified using the internet. A detailed study recognized that Javelin Strategy and Research reported crime of 15.4 billion USD in 2015 and 16 billion USD in 2016, respectively [14]. A comprehensive survey revealed that 4,248 US adults had been found to analyze the different levels of crime severity. The result analysis proved that there is 41% online harassment and 66% online mistreatment [15]. The study also detected that 18% of the samples prevailed in severe online abuse comprising physical intimidations, sexual persecution, and annoyance.

The internet has become a venue for illegal activities that provide an endless supply of potential victims. The virtual world observed crimes committed against youth. A survey report involved 1,024UK young people where 28% experienced stressful events such as harassment and annoyance over network profiles [16]. The Internet Watch Foundation identified 57,000 web pages covering sexually manipulative imagery. Also, the survey reported that 34% experienced diverse types of online mistreatment, while 12% accounted to online victimization [17].

1.4 Motivation

To address the issues of digital forensics, researchers and investigators promoted integrating behavioral techniques. The other disciplines [18] investigate the inductive approaches to classify the taxonomies of cybercrimes. The indicator and generality are dependent on the creation of computer-specific crimes. However, a comprehensive approach is inadequate to investigate the practical applicability and specificity of computer-based criminalities. This chapter includes an extensive examination to investigate digital crimes incorporating behavioral and emotional analysis. A detailed review considered the development of the investigation process to show the integral aspects of behavioral analysis. Thus, this chapter demonstrates the effectiveness of digital forensic study that includes investigation, analysis, and clarification of the digital proof of computer-based criminalities.

1.5 Objectives

In this chapter, a crucial aspect of the digital investigation process is considered to constitute the evidence of systematic interrogation, analysis, and interpretation. Moreover, it is aimed to explore digital data and to investigate the types of computer-based digital crimes. The main objectives are as follows:

1. Examine the relevance and usability of digital investigation to explore the evidence of digital crimes.
2. Observe the facilities of digital forensics to understand the behavioral and emotional analysis.
3. Study the current issue and the evidential aspects of digital policies and enforcement laws to perceive the digital investigation.
4. Integrate the precedence of investigation tools to incorporate the objectives of digital evidence including the broad aspects of online crimes.

2. Background

This section discusses the collected reviews to summarize the criminal interpretation. The characterization of criminal interpretation and development histories are covered to review and appraise the previous criminal representation including the general terms of social behavioral analysis (SBA) and subset of behavioral evidence analysis (S-BEA) [19].

2.1 Criminal recording

In a criminal investigation, criminal recording is used as a forensic technique to investigate the phases such as analysis, evaluation, and interpretation of the forcible evidence. The nature of the corruption scene is recurred to produce a statistical profile that examines the behavioral characteristics of a lawbreaker contrary to the features [20]. The objective is to create a demographic record and behavioral characteristics of a lawbreaker contrary to the committed crimes [21]. The criminal recording practices three extensive forms of inferential perceptive to create a subjective profile: induction form, deduction form, and abduction form.

Induction form applies behavioral analysis and emotional data from convicted offenders and relies on criminal databases to classify the generalized pattern and persona attributes.

Deductive form relates to case-based inquiries and analyzes the case evidence in the subject to focus on the specific behavioral pattern and traits, using a criminal record to characterize the offender.

Abduction form involves systematic inferences to explain the available evidence to create an inadequate set of opinions that attempts to draw the predicted sets of conclusion [22].

2.2 Social behavioral analysis (SBA)

Criminal recording refers to common terms, such as interactive analysis, behavioral recording, crime scene recording, offender record, psychological record, investigative record, and criminal analysis [23]. Due to the nonexistence of uniformity, the substitutable and unpredictable considers a general technique to investigate the criminal recording. Fig. 6.2 shows the subsets of criminal recording.

2.3 Subset of behavioral evidence analysis (S-BEA)

S-BEA is an inferential subset that uses a case-based strategy to analyze criminal evidence that primarily focuses on certain behavioral and personality traits to deduct the offender's characteristics.

FIGURE 6.2

Subsets of criminal recording.

2.3.1 Equivocal forensic analysis

Casey et al. [24] performed an ambivalent forensic analysis that refers to the course of systematic assessment including the case evidence:

- An exhaustive investigation, analysis, and assessment of criminal evidence is employed to infer cognitive thoughts, logical reasoning and analysis.
- A possible clarification interprets the evidence to determine the credible importance of the criminal data.
- A critical assessment considers the case assumptions to review the digital evidence.

2.3.2 Victimology

Victimology studies illegal investigations to explain criminal opportunities and involves a systematic study of adverse characteristics, regular procedures, and routine activities to identify victim particularities.

2.3.3 Crime scene

Digital crime permits criminals to use multiple locations to confuse investigators to avoid the detection of virtual location in cyber-space.

2.3.4 Lawbreaker characteristics

This tool is used to commit the crime that signifies a routine strategy to represent the offender. The detection measurement avoids substantial evidence of measures to blot out the offender's behavior. It offers a technical indication to investigate personality traits at a sophisticated level.

2.3.5 Principles of subset of behavioral evidence analysis

S-BEA derives basic principles identified by behavioral analysis to summarize technical possibilities:

- *Uniqueness:* Two identical cases may acquire similar characteristics to investigate the digital crimes.
- *Separation:* A separate investigation may be chosen to analyze the digital activities.
- *Dynamic behaviors:* An individual behavior may be analyzed to infer the conscious activities of digital evidence.
- *Multiple determination:* A single behavior may analyze the evidence of behavioral pattern to investigate the criminal case.
- *Reliability:* Digital forensics may analyze the tangible evidence and reasoning to investigate the criminal activities.
- *Behavioral variant:* Digital evidence may involve criminal intentions to examine the nature of the criminal offense.

3. Digital forensics

Digital forensics is a branch of forensic science that is called computer forensics and forensic computing to uncover the representation of electronic data. It performs incidental evidence to commit a digital crime that investigates the connective crime as a tool to determine the nature of evidence. The goal of the digital investigation is to analyze the conventional use of criminal inquiry that commits several criminal activities. The analytical methods use a conservative crime that is nowadays incorporating digital forensics to deal with scientific proven technique involving digital identification, data preservation, collecting criminal data, validating the criminal activities, analyzing the evidence, interpreting the criminal records, documenting the criminal cases, and presenting the digital evidence. The restoration of an illegal event recovers the digital evidence to facilitate the incrimination or exoneration of suspects and to abide by criminal prosecution.

The electronic evidence must recover forensically to preserve digital integrity, preserving the criminal opinion, destruction, and corruption inadmissible with the law. This is entailed to improve the procedural process that handles the digital evidence and policies effectively. The digital investigator must ensure device integrity to investigate the recovered evidence that applies prosecution methods to assure the trustworthiness and productivity in the support of digital crime. Law enforcement or lawful schemes challenge the successful prosecution of digital evidence that includes legal prosecution, admissibility of digital tools, techniques, privacy, and ethical issue. The forensic requirements include education, qualifications, and training to address the digital evidences, semantic inequalities in digital forensics, lack of unified formal representation of development factor (DF), domain knowledge, the nonexistence of criminal information, and legal authorization.

The functional challenges contain prevalence recognition, reaction and preclusion, privation of consistent process and techniques, substantial labor-intensive intercession and investigation, digital organizations, and conviction of audit trails. The conventional method may no longer be practical to investigate digital crimes, so criminal activity may involve new approaches to address the existing challenges.

4. Investigation models in digital forensic

Digital evidence fairly compares the other forensic fields, with uninterrupted hard work to improve the existing methods and standards that regulate the forensics investigation process. The nonexistence of regulation and model inconsistencies may result in imperfect evidence, inaccuracy of interpretation, and limited admissibility in the court of law [25]. A standard forensic model may improve the technical accuracy of the digital process that facilitates the virtue of digital applicability and research direction. The technical researchers and experts develop a standard evidence model to highlight the critique of digital forensics.

4.1 Model 1: crime scene investigation

The digital crime considers four investigation phases: secure and scene evaluation, criminal documentation, collecting the evidence, and transporting the digital evidence. It refers to the electronic evidence to find proper methods to hold the collection of digital information, showing specific importance to address the crime scene.

4.2 Model 2: abstract digital forensics

Reith et al. [26] claimed trials of a digital forensic model to generalize the use of existing methodologies that include digital identification, preparing the evidence, developing strategies, collecting evidence, examining a model, analyzing data, presenting digital evidence, and returning the belief of evidence.

4.3 Model 3: integrated investigation process

Carrier et al. [27] integrated the investigation process to combine physical and digital crime that involves five groups such as promptness, placement, physical crime scene, digital crime scene, and investigation review. Contrasting with previous investigation models, the integrated integration highlighted the significance of a physical crime scene in digital form. It delivered the use of primary and secondary crimes respectively.

4.4 Model 4: hierarchical objective-based (HOB) framework

Beebe et al. [28] suggested the HOB framework to investigate digital evidence that uses a multitier process to provide the details of particularity and practicality. The previous works compared the first tier of an objective-based framework to generate the cooperation process between dissimilar technical perspectives. It includes six phases: preparing the digital evidence, creating the incident response, collecting the digital data, analyzing the evidence, presenting the digital findings, and returning the incidental closure.

4.5 Model 5: Cohen's digital forensics

Cohen et al. [29] addressed the legal issues of digital evidence to represent the forensic elements such as investigation, clarification, ascription and reconstruction. These elements are described in general form to exhibit the use of tangible cases.

4.6 Model 6: systematic forensic investigation

Agarwal et al. [30] presented a systematic investigation model to include organized preparation, securing criminal scenes, recognizing the review materials, documenting the crime scene, communicating the evidence, collecting the criminal data, preserving the crime scene, examining the criminal data, analyzing the crime, presenting the criminal defense, and results of the crime scene.

4.7 Model 7: harmonized forensic investigation

Valjarevic et al. [31] presented a forensic investigation method to harmonize existing models such as iterative and multilayered. These models include detecting the incidental request, incidental response, planning the investigation, preparing the criminal scene, documenting the incidental scene, identifying the potential evidence, collecting the evidence, transporting the digital evidence, storing the digital data, analyzing the evidence, and concluding the investigation.

4.8 Model 8: integrated forensic investigation

Montasari et al. [32] claimed to address the essential requirements of the investigation process that include promptness, identifying the crime scene, incidental response, collecting the crime scene, examining the digital evidence, analyzing the crime scene, presenting the digital evidence, and concluding the incidental closures.

4.9 Model 9: Mir's forensic model

Mir et al. [33] studied a model to limit the scope of three investigation processes: acquiring the digital evidence, insubstantial evidence, and examining the digital information. Moreover, this model encompasses planning the process, identifying the evidence, collecting the digital data, exploring the information gain, storing and transporting the information, examining the evidence, analyzing the digital grounds, creating the proofs, archiving the evidence, and presenting the evidence.

4.10 Model 10: general limitation

Most of the previous models use a single tier that focuses on high-level investigation to provide the investigation principles to suggest some additional specific steps including constituting the evidence and identifying the guidelines. These steps are recommended to offer sufficient detail to investigate digital evidence.

5. Integrating behavioral analysis

The published researches struggled to integrate behavioral analysis within the analysis of interactive crimes that have ownership and distribution of sexually exploitative imagery of children. Moreover, they have cyberstalkers to address digital forensics profiling methodology and behavioral analysis.

5.1 Cyberstalker: forensic methodology

Silde et al. [34] developed a forensic methodology to incorporate the elements of behavioral analysis that includes standard forensic investigation to constitute discovering the digital evidence, examining the digital data, and analyzing the evidence. The evaluation method assesses the utility model to simulate the predefined set of cyber activities. The researchers examine the victim's evidence and criminal technologies.

5.2 Roger's behavioral analysis

Roger [23] analyzed the field of digital forensics to address the engineering principles that focus on the investigation process concerning data collection and examination. This model includes case classification, analyzing the contextual data, collecting the digital data, visualizing the digital evidence, and concluding the opinion. The classification case identifies the investigation cases such as scams, cyberstalking, data theft. The contextual analysis understands digital circumstances to deliver the visions of relevant evidence. The data collection works with the interactive analyst to search the relevant data in the preparation of digital evidence. The visualization focuses on guiding the incidence analysis that identifies the interpretation patterns. The conclusion/opinion concerns the production of the final report that reports the analytical questions to start the inquiry.

6. Sexually exploitative child imagery

Online distribution and delivery identify the offenders involved in the production or possession of any pictorial portrayal of teenagers, i.e., under a certain age. The online distribution may include the digital transformation, i.e., pornographic material, that uses digital devices to project and produce the virtual streaming of erotic exploitation.

6.1 Online prevalence

A detailed study is conducted in the dissemination of online possession that uses peer-to-peer (P2P) to share and receive online materials. The internet is largely credited to protect collective data that reveals online crimes. The offenders are increasingly using new techniques to masquerade from criminal authorities.

6.2 Characterize the sexual victims

Content analysis showed that the preponderances of victims are females, i.e., 89%. It also exhibited that 53% of the youngsters seem to be under the age of 10. A potential explanation demands a new trend of *"self-protected"* content surrounding adolescents.

6.3 Characterize the sexual offenders

The empirical research examines the behavior, demographics, and emotional characteristics of sexual offenders. The sexual offenders should face criminal justice to start treatment systems under the psychosomatic characteristics and incentives. These opportunities will make offenders understand the consequences, such as prison and charges. Also, the detailed study realized the importance of law enforcement agencies to convict offenders for sexual offenses, both online and offline.

7. Offender typologies and theories

The motivating factors involve offender collection that focus on erotic interest to relate with nonsexual obsessive use.

7.1 Offender motivation

The two main offender motivations are erotic interest that provides the equivalent erotic satisfaction provided by rendezvous with the obsessive materials. Quayle et al. [35] recognized erotic arousal, gaining erotic pleasure, avoiding real-life issues, social relationships, emotional therapy, and manifesting addictive properties. The study findings are based on semistructured interviews that condemned the offenders for assessing online downloads and file ownership. The discussions primarily were instituted to explain open-ended queries that recorded, implied, and examined the patterns of motivation factors.

7.2 Offender typologies

Researchers attempted to categorize offender characteristics using the nature of behavioral and emotional features. These key features include criminal incentives to possess the level of practical skills and level of participation in the dissemination of online communities, whereby a proper countermeasure should be taken to circumvent discovery and evolution of sexual exploitation.

- *Browser* mentions individual views of sexual exploitation to determine the examination of the contiguous facts and digital proof.
- *Private fantasy* discovers the representation of erotic fantasy to gain private access.
- *Trawler* searches the online erotic interest to analyze security measures that do not mask behavior interest.
- *Nonsecure amasser* practices the websites to restrict network resources including user, passwords, and encryption to purchase, trade, or transfer erotic contents.
- *The secure collector* obtains secure groups and online clusters that exhibit the resilient desires to create a large collection of online offenders.
- *Online groomer* pursues online contact to form a relationship with a minor, which gradually builds up the development of online or physical erotic contact.
- *Physical abuser* involves erotic exploitation that uses online contacts to distribute or share the online materials.
- *Producers* commit abuse against physical minors by record files and sharing the contents with others.
- *Distributors* may not have any erotic interest in youngsters but are mainly interested in financial gain by disseminating the online contents.

7.3 Cyberstalkers

The term usually denotes the use of internet technologies to follow or pursue other activities. It involves analyzing the inappropriate material that may harm the targeted victim to hide the imperceptible technologies. The studies are conducted to identify three main aspects: stalking email services, stalking internet service, and stalking computer facilities.

8. Research methodologies

Saunder et al. [36] classified the research methodology into five-layers (research onion) that describe the perspectives of general research paradigms. The outmost layer signifies the inferences for the study project. It supports identifying a project model based on positivist, informative, perilous, realistic traits. The second layer mentions the procedural choices to provide precise results including qualitative and quantitative methodology. The third layer adapts the specific approach to respond to the questions, i.e., action survey, whereas the fourth deliberates the horizon to address the difficulties, i.e., cross-sectional or longitudinal. Lastly, the research core highlights the procedures and techniques undertaken to analyze the behavioral research (Fig. 6.3).

8.1 Research paradigm

Research standards state the beliefs of the investigator that evolve and corroborate scientific knowledge to embark on the growth of knowledge impacts and outline the investigation design.

8.2 Methodology

Qualitative research delivers an insightful phenomenon to perform detailed studies on a trivial model group to discover and recognize theoretical concepts. Moreover, it

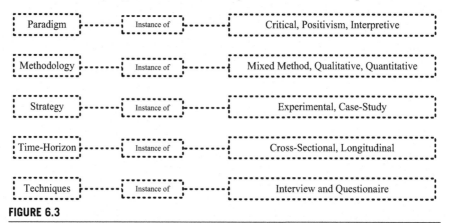

FIGURE 6.3

Hierarchic representation of research methodology.

guides the construction of the hypothesis. Qualitative research conducts some strategies such as conferences, case revisions, grounded model, and ethnography. The composed data is analyzed and thematic study and content study are used to categorize the themes or patterns in the cohort of hypotheses.

8.3 Research strategy

This strategy uses unambiguous data collection and investigation methods to discourse the queries of the investigation and has several strategies to employ forensic research. The major approaches are grounded theory, case studies, action research, experiments, and quasiexperimental.

9. Conclusion

This chapter has extensively studied the standard exercise that covers major aspects of digital forensics integrated with behavioral analysis. It has included four stages of SBA for the examination of digital evidence. This case study showed that a general approach utilizes computer-based social crimes to investigate the social evidence. It uses the investigation approach to provide rational instructions that identify the relevant evidence to escalate the investigation process. Systematic behavior and offender interpretation facilitate an in-depth analysis to eliminate online suspects. This implication shows the practical abilities to relate the behavioral analysis within the interrogative process of digital forensics, which deals with a specific category of computer-assisted criminal activities. In the past, it has been hypothesized that behavioral analysis includes four stages to investigate forensic processes. However, it has not been verified empirically. Thus, this chapter gathered the knowledge evidence to analyze forensic investigations and provided insights into the investigation process. The study work has provided adequate strategies to apply the investigation process within digital forensics. In the future, adequate educational practice and training will be provided to examine the investigation process.

References

[1] V. Baryamureeba, F. Tushabe, The enhanced digital investigation process model, in: Proceedings of the Fourth Digital Forensic Research Workshop, August 2004, pp. 1–9.

[2] H. Arshad, A.B. Jantan, O.I. Abiodun, Digital forensics: review of issues in scientific validation of digital evidence, J. Inf. Process. Syst. 14 (2) (2018).

[3] H. Arshad, A. Jantan, G.K. Hoon, A.S. Butt, A multilayered semantic framework for integrated forensic acquisition on social media, Digit. Invest. 29 (2019) 147–158.

[4] S. Casale, G. Fioravanti, Satisfying needs through social networking sites: a pathway towards problematic internet use for socially anxious people? Addict. Behav. Rep. 1 (2015) 34–39.

[5] D. Dutta, C. Tazivazvino, S. Das, B. Tripathy, Social Internet of Things (SIoT): transforming smart object to social object, in: NCMAC 2015 Conference Proceedings, 2015.

[6] N. Malamuth, D. Linz, R. Weber, The Internet and Aggression: Motivation, Disinhibitory, and Opportunity Aspects, Oxford University Press, 2013.

[7] A. Rashid, A. Baron, P. Rayson, C. May-Chahal, P. Greenwood, J. Walkerdine, Who am I? analyzing digital personas in cybercrime investigations, Computer 46 (4) (2013) 54–61.

[8] S. Vahdati, N. Yasini, Factors affecting internet frauds in private sector: a case study in cyberspace surveillance and scam monitoring agency of Iran, Comput. Hum. Behav. 51 (2015) 180–187.

[9] H. Chen, C.E. Beaudoin, T. Hong, Protecting oneself online: the effects of negative privacy experiences on privacy-protective behaviors, J. Mass Commun. Q. 93 (2) (2016) 409–429.

[10] H. Sampasa-Kanyinga, H.A. Hamilton, Use of social networking sites and risk of cyberbullying victimization: a population-level study of adolescents, Cyberpsychol. Behav. Soc. Netw. 18 (12) (2015) 704–710.

[11] H. Chen, C.E. Beaudoin, T. Hong, Securing online privacy: an empirical test on Internet scam victimization, online privacy concerns, and privacy protection behaviors, Comput. Hum. Behav. 70 (2017) 291–302.

[12] T.J. Holt, Situating the problem of cybercrime in a multidisciplinary context, in: Cybercrime through an Interdisciplinary Lens, Routledge, 2016, pp. 15–28.

[13] D.S. Wall, Cyber Crime: The Transformation of Crime in the Information Age, Polity, UK, 2007.

[14] A. Pascual, K. Marchini, S. Miller, Identity Fraud: Securing the Connected Life, vol. 20, 2017, p. 2018. Retrieved on January.

[15] Pew Research Center, Online Harassment 2017, 2017, p. 85.

[16] C. Lilley, R. Ball, H. Vernon, The Experiences of 11–16 Year Olds on Social Networking Sites, National Society for the Prevention of Cruelty to Children (NSPCC), 2014. Retried from, www.nspcc.org.uk/Inform/resourcesfor-professionals/onlinesafety/11-16-social-networkingreport_wdf101574.pdf,10.

[17] G.N. Samy, B. Shanmugam, N. Maarop, P. Magalingam, S. Perumal, S.H. Albakri, Digital forensic challenges in the cloud computing environment, in: International Conference of Reliable Information and Communication Technology, Springer, Cham, April 2017, pp. 669–676.

[18] R. Bryant, Policing Digital Crime, Routledge, 2016.

[19] B.E. Turvey, Forensic Victimology: Examining Violent Crime Victims in Investigative and Legal Contexts, 2 ed, Academic Press, 2014, p. 656.

[20] P. Ainsworth, Offender Profiling Crime Analysis, Willan, 2013.

[21] M. Yar, K.F. Steinmetz, Cybercrime and Society, SAGE Publications Limited, 2019.

[22] I.A. Fahsing, K. Ask, In search of indicators of detective aptitude: police recruits' logical reasoning and ability to generate investigative hypotheses, J. Police Crim. Psychol. 33 (1) (2018) 21–34.

[23] M.K. Rogers, Psychological profiling as an investigative tool for digital forensics, Digit. Forensics: Threat. & Best Pract. 45 (2015).

[24] E. Casey, A. Blitz, C. Steuart, Digital Evidence and Computer Crime, Academic Press, 2014.

[25] X. Du, N.-A. Le-Khac, M. Scanlon, Evaluation of Digital Forensic Process Models with Respect to Digital Forensics as a Service, 2017 arXiv preprint arXiv:170801730.

[26] M. Reith, C. Carr, G. Gunsch, An examination of digital forensic models, Int. J. Digit. Evid. 1 (3) (2002) 1−28.

[27] B. Carrier, E.H. Spafford, Getting physical with the digital investigation process, Int. J. Digit. Evid. 2 (2) (2003) 1−20.

[28] N.L. Beebe, J.G. Clark, A hierarchical, objectives-based framework for the digital investigations process, Digit. Invest. 2 (2) (2005) 147−167.

[29] F. Cohen, Toward a science of digital forensic evidence examination, Adv. Digit. Forensics VI (2010) 17−35.

[30] A. Agarwal, M. Gupta, S. Gupta, S.C. Gupta, Systematic digital forensic investigation model, Int. J. Comput. Sci. Secur. 5 (1) (2011) 118−131.

[31] A. Valjarevic, H.S. Venter, Harmonised Digital Forensic Investigation Process Model, Information Security for South Africa, Johannesburg, 2012, pp. 1−10.

[32] R. Montasari, P. Peltola, D. Evans, Integrated computer forensics investigation process model (ICFIPM) for computer crime investigations, in: International Conference on Global Security, Safety, and Sustainability, Springer, 2015, pp. 83−95.

[33] S.S. Mir, U. Shoaib, M.S. Sarfraz, Analysis of digital forensic investigation models, Int. J. Comput. Sci. Inf. Secur. 14 (11) (2016) 292.

[34] A. Silde, O. Angelopoulou, A digital forensics profiling methodology for the cyberstalker, in: Intelligent Networking and Collaborative Systems (INCoS), 2014 International Conference on: IEEE, 2014, pp. 445−450.

[35] E. Quayle, M. Taylor, Child pornography and the Internet: perpetuating a cycle of abuse, Deviant Behav. 23 (4) (2002) 331−361.

[36] M. Saunders, Research Methods for Business Students, 7 ed., Pearson Education, India, 2015, p. 768.

Recommender systems: security threats and mechanisms

7

Satya Keerthi Gorripati[1], Anupama Angadi[2], Pedada Saraswathi[3]

[1]*Computer Science and Engineering, Gayatri Vidya Parishad College of Engineering (Autonomous), Visakhapatnam, Andhra Pradesh, India;* [2]*Computer Science and Engineering, Raghu Engineering College (Autonomous), Dakamarri, Viskhapatnam, Andhra Pradesh, India;* [3]*Department of Computer Science and Engineering, GITAM Institute of Technology, Visakhapatnam, Andhra Pradesh, India*

1. Introduction

Recommender systems (RSs) have been an attractive field since the early 1990s. They were applied in many fields in the movies domain (e.g., Netflix, Filmaffinity, and MovieLens), web pages (e.g., Quora, Stackoverflow), music tracks (e.g., Wetracker, Beepbox), e-commerce (e.g., Amazon, Flipkart, and Myntra), and social media (e.g., Facebook, Twitter, and Instagram) to avoid information overload and customize services. Information overload arises when there is difficulty in understanding an issue and failure to make effective decisions when one has very large data about that issue. RSs alone do not transform this data into information; they are just an individual element of the information conduit. Data preprocessing tasks convert those issues of data into valuable information, and then the RS supports to filter that information into the most significant, from which a human can mine knowledge and take action. Note that through information filtering, RSs [1] provide efficient customer service.

The primary components of RSs are (i) information filtering, (ii) user behavior, (iii) suggesting information, (iv) end-user goals, and (v) security. Information filtering does not infer the deletion of information but rather prioritizes information based on the users' interests. An RS supports the users to take proper choices by exploiting their profiles and predicting or suggesting new choices. The end-user goal is a final form for which an RS strives. Finally, an online RS requires the users' profiles to predict their needs, and this data can point to the users' uniqueness (e.g., e-mail, SSNs), as well as their browsing and purchase history, suggesting user likes and dislikes. Owing to the discrepancy in sensitivity among the users' data, it is ineffective to implement a universal security approach without losing the property of accuracy [2].

Security in IoT Social Networks. https://doi.org/10.1016/B978-0-12-821599-9.00007-8
Copyright © 2021 Elsevier Inc. All rights reserved.

Acronym	Abbreviation	Acronym	Abbreviation
RS	Recommender systems	CF	Collaborative filtering
TF	Term frequency	IBCF	Item-based CF
IDF	Inverse document frequency	UBCF	User-based CF
CBRS	Content-based RS		

1.1 Motivation

RSs are powerful tools helping online users to tame information overload by providing personalized recommendations on various types of products and services. Not surprisingly, users, who are looking for the next book to read or a digital camera to buy, are overwhelmed by the quantity of information. Hence, they may find it difficult to filter out the irrelevant information and choose the most suitable products without some assistance tailored to the specific needs and knowledge of the user.

To address these problems, RSs have been proposed and now are largely diffused. These are intelligent applications that are able to identify and suggest products, information, or services that best fit the user's needs and preferences. Social RSs are an active and growing research area. They raise interesting and difficult problems related to the improvement of the existing approaches to exploit and model new privacy solutions. This chapter evaluates how users' delicate and private information is typically gathered and calculated when producing recommendations. We found specific literature on privacy salvation from numerous features, namely, trusted broadcast and types of attacks and encryption methods. However, current methods not suitable for distributed recommender applications. How to secure user ratings, neighboring data, and aggregated information without sacrificing accuracy on trusted RSs is still an exposed topic.

2. Features of recommender systems

Typically, the recommendation process starts when the users show their views/preferences by rating items, as shown in Fig. 7.1. The necessary postulation of RSs is that if users P and Q prefer n corresponding products, it is likely they possess comparable behaviors (e.g., viewing, purchasing, and listening), so they will act or turn to further items correspondingly. RS methods use a data store of the users' preferences to guess supplementary products or topics a novel user might prefer. In a classic RS situation, there is a catalog of m users $\{u_1, u_2, ... u_m\}$ and catalog of n items $\{i_1, i_2, ... i_n\}$, and every user u_i has a record of preferred items, Ru_i, which the user has chosen, by which their favorites have been concluded. The ratings can be on a 1 to 5 scale.

In the initial days, most of the first-generation investigators viewed RSs as two-dimensional *User X Item* features. Due to restricted computational resources and

FIGURE 7.1

General prototype of a recommender.

deficiency of knowledge, they made RSs as a matrix of user and item features. Later, with the success of machine learning and deep learning, investigators found that additional features can suggestively enhance RSs. For example, demographic profiling (e.g., gender, living place, and age), castoff cold start difficulty [3], and progressive information can be functional to seize the users' flowing interests, and locality-based and social network information can be used not only to progress RSs but also to create a modern RS [4,5]. Much popular existing literature on RS adopts the intake as the users' rating data, which varies with the type of the source [6]. The following are some of the sources and types of information, as shown in Fig. 7.2 and Table 7.1.

The users' obvious feedback on items is the most fundamental method to build a *User X Item* matrix. For example, the users rate a taxi driver, a movie, or a guest house with a score on a 1 to 5 scale, and scale 0 denotes the fact that the item is not rated. But in actuality, very few items can obtain explicit feedback [7]. So gathering implicit feedback, such as product views, online news clicks, history of watched videos, browsing tourist places, and so on, is essential to build a *User X Item* matrix. In case of implicit feedback, the entry in the matrix is always the number of visits, and investigators attempt to enumerate that into numeric, usually denoted as shown:

$$f: UxI^{mxnxd} \rightarrow R^{mxn} \tag{7.1}$$

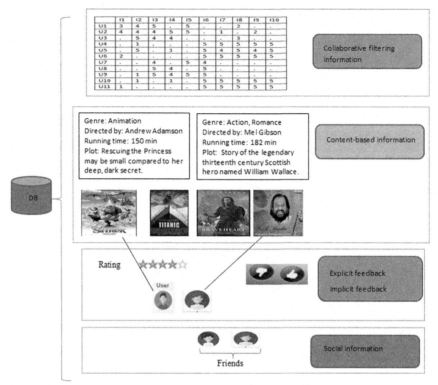

FIGURE 7.2

Sources of information for database.

Table 7.1 Sources of information for RS.

	News	Viewing sites	Video review	Music	Books	Social network
Explicit rating		√	√	√	√	√
Implicit rating	√	√	√	√	√	
Time	√	√	√			
Locality	√	√	√	√		√
Content	√	√	√	√	√	√
Sentiment			√	√		
Quotation		√	√	√		√

where d represents the features of RS, and $d = 1$ specifies two-dimensional features. If extra features, like locality, time, and social connections, are involved, $d - 1$ is the additional number of supplementary features [4]. The function f is the recommendation principal that predicts the user's preferences.

A typical view of a *User X Item* matrix is shown in Fig. 7.2, where every numeric entry denotes the feedback of the user on a specific item. However, this origin has lots of sparse elements, proposing that there are anonymous preferences (e.g., P/R). The symbol "?" in R denotes an unknown rating. The task of an RS is to predict these unknown ratings.

Concerning the outputs of an RS, there are typically two possibilities that exist: either distinct or $Top - N$ prediction. Distinct prediction outputs a specific prediction for an unrated item, whereas $Top - N$ prediction outputs a prediction for N unrated items.

3. Common classification of recommender systems

3.1 Content-based approach

Most prevailing CBRS emphasis suggesting items with textual data such as blogs, news articles, and booklets [8]. The information in these structures is typically labeled with crucial words. The relevance of a crucial word to an article is frequently considered by TF-IDF weight [9,10] as shown in Fig. 7.3. TF is the worth of a crucial word to an article or signifies the occurrence of the crucial word in an article, while the IDF worth of a crucial word is described as the inverse document frequency of the crucial word, usually denoted as shown in Eq. (7.2) [6]:

$$f : (PRofile, Content) \rightarrow r_{u,i} \tag{7.2}$$

where $r_{u,i}$ is the measure of how much the user u likes the item i. Profile and content denote the users' profile and data specification of the item.

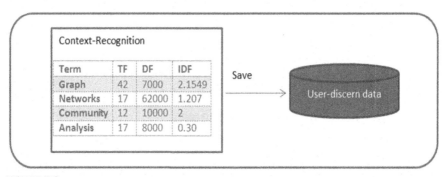

FIGURE 7.3

Item profile of TF-IDF model.

3.2 Collaborative filtering (CF) approach

CF is the popularly used method to build RSs, which can be further categorized into memory-based and model-based techniques [11] In memory-based technique, the preferences (e.g., *User X Item*) matrix is considered an input and predicts the user preferences rightly by revealing uncovered and complex patterns from the users' past behavioral data and mentions preferred items to the target the user from a matched set of the users with similar interests in the future [12]. The CF approach assumes that if the user's choices have coincided with one another previously, they are further more probable to coincide with one another subsequently than to choose randomly selected users [13]. The memory-based approach can be further distributed into IBCF and UBCF methods, usually denoted as shown [6]:

$$f: (User, Item) \rightarrow r_{u,i} \tag{7.3}$$

For instance, consider the user rating matrix of 11 users and 15 items to predict the rating score of item i_4 for user u_1 with KNN CF algorithm oriented to users. To compute the prediction, first calculate the similarity between user u_1 with other users using a similarity metric, as shown in Table 7.2. The computed similarity values are listed under Pearson similarity column. The highest similarity value users (u_2 and u_3) are selected as neighbors for the target user u_1, and by aggregating these neighbors ratings, e.g., u_2 given rating 5 and u_3 given 5 rating to item i_4, the predicted value is equal to 4.5 ($(5 + 4)/2$). This process continues for every unrated item of the target user to build his/her complete profile.

Consider the user—item rating matrix of 11 users and 15 items to predict the rating score of item i_4 for the user u_1 with KNN CF algorithm oriented to items. To compute the prediction, first, we find the similarity between item i_4 with other items using a similarity metric, as shown in Table 7.3. The computed similarity values are listed under the similarity row. The highest similarity value items (i_3 and i_5) are selected as neighbors for the target item i_4 and by aggregating these neighbors ratings (i_3 given rating 5 and i_5 given 5 rating to user u_1), the predicted value is equal to 5.00.

3.3 Hybrid approach

It combines multiple RS techniques, like combining collaborative and content-based approaches, to produce the output [14].

4. Sources of recommender systems

Based on the source of the data the RS is divided into collaborative, content-, and knowledge-based recommendations.

Collaborative sources: These sources rely on numerous notions to designate the problem field and the specific necessities placed on the system. In the CF technique,

Table 7.2 Users rating data and its similarity using UBCF.

Pearson similarity		1	2	3	4	5	6	7	8	9	10	11	12	13	14	15
1	U1	3	4	5	—	5	—	—	2	—	—	—	—	—	—	—
0.6485	U2	4	4	4	5	5	—	1	—	2	—	1	—	—	—	3
0.7218	U3	—	5	4	4	5	—	—	3	—	—	—	2	—	—	1
−0.284	U4	—	1	—	—	—	5	5	5	5	5	—	—	—	—	—
−0.1464	U5	2	5	—	1	—	5	4	5	4	5	—	2	—	—	—
−0.2979	U6	—	—	—	—	—	5	5	5	5	5	—	1	5	4	5
−0.4641	U7	—	—	—	—	—	—	—	—	—	—	4	5	5	5	4
−0.4641	U8	—	—	—	—	—	—	—	—	—	—	5	4	5	5	5
−0.4301	U9	—	—	—	—	—	—	—	—	—	—	5	4	5	—	—
−0.3078	U10	—	1	—	1	—	5	5	5	5	5	—	—	—	—	—
−0.3483	U11	1	—	—	—	—	5	5	5	5	5	—	1	—	—	1

Table 7.3 Users rating data and its similarity using IBCF.

Similarity	1.0	0.29	0.59	0.38	0.63	−0.2	−0.1	−0.1	−0.01	−0.2	−0.3	−0.5	−0.4	−0.4	−0.1
	1	2	3	4	5	6	7	8	9	10	11	12	13	14	15
U1	1	2	3	4	5	—	—	2	—	—	—	—	—	—	—
U2	3	4	5	—	5	—	1	—	2	—	1	—	—	—	3
U3	4	4	4	5	5	—	—	3	—	—	—	2	—	—	1
U4	—	5	4	4	5	5	5	5	5	5	—	—	—	—	—
U5	—	1	—	1	—	5	4	5	4	5	—	2	—	—	—
U6	—	5	—	—	—	5	5	5	5	5	—	1	—	—	—
U7	2	—	2	—	—	—	—	—	—	—	4	5	5	4	5
U8	—	—	—	—	—	—	—	—	—	—	5	4	5	5	4
U9	—	1	—	—	—	—	—	—	—	—	5	4	5	5	5
U10	—	1	—	1	—	5	5	5	5	5	—	—	—	—	—
U11	1	—	—	—	—	5	5	5	5	5	—	1	—	—	1

the users assist each other to identify fascinating items. Every user rates a subset of items with some scores, so CF does not cover the content description of the items [9]. In this context, the user rating shows his/her willingness or interest toward that item. The CF approach assumes that if the users have agreed with one another previously, they are further probable to agree with one another subsequently over choosing randomly selected users. The memory-based approach can be further separated into UBCF methods and IBCF methods, as shown in Fig. 7.4. In UBCF, the recommender predicts unknown item ratings through aggregating the ratings provided by other users, weighted with the similarity between the users [15]. In IBCF, the recommender predicts unknown item ratings through aggregating the ratings provided by other items, weighted with the similarity between items [16]. Usually, the preference matrix is mostly sparse, proposing that there are a large number of unrated items.

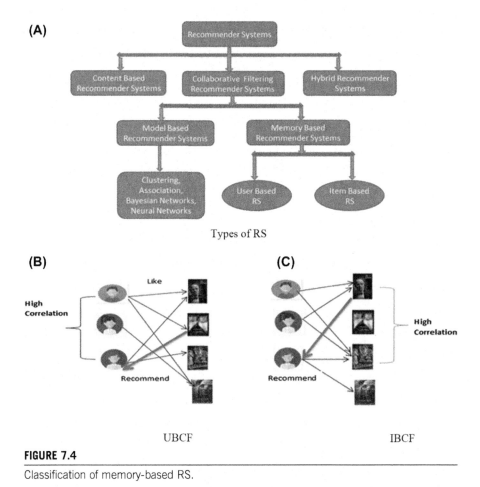

FIGURE 7.4

Classification of memory-based RS.

CF is the most familiar type of RS in the literature. Many other algorithms use CF, since examining large *User X Item* matrices is a tedious task for traditional databases, and classifying them is still a piece of work where humans surpass machines. In this case, perceiving consumer behavior and not examining larger matrices seems a reasonable and convenient way to produce accurate recommendations.

Unfortunately for the CF, the activity of acquiring the users' history may be difficult. This data can use either implicit or explicit methods. Some of the explicit samples are (i) querying the consumer to rate the ambiance of a hostel room, (ii) asking the user to rank comfort of the service, or (iii) asking the user to pick his favorite items. Some of the implicit samples include (i) perceiving the online purchased patterns of the users, (ii) noticing the frequently observed items in browsing history, or (iii) notifying the user's interest by pressing thumbs up or thumbs down.

The most commonly used algorithm in CF is the k-nearest neighborhood method [17]. In a social network, a certain neighborhood of the users with common taste can be computed by using Pearson's correlation [18,19]:

$$sim(p, q) = \frac{\sum\limits_{l \in ADJ_{a,u}} (r_l(p) - \bar{r}(p))(r_l(q) - \bar{r}(q))}{\sqrt{\sum\limits_{l \in ADJ_{a,u}} (r_l(p) - \bar{r}(p))^2} \sqrt{\sum\limits_{l \in ADJ_{a,u}} (r_l(q) - \bar{r}(q))^2}} \qquad (7.4)$$

where $sim(p, q)$ is the commonality between the users p, q, $\bar{r}(p)$ is the users' average rating, $r_l(p)$ is the users' rating to an item p, and $pred(a, p)$ the predicted rating of item p.

We concentrate our efforts on defining one that is (i) nimble, (ii) disseminated, and (iii) secure. RSs and e-business are tightly correlated fields. An RS becomes useful if it happens quickly, and the participation is enhanced if recommendations are both nimble and accurate [18]. Disseminated RSs are contemplated as more secure than centralized RSs because they avoid the modeling of a single database with all information, thereby preventing standard centralized security attacks [20] Furthermore, modeling a disseminated system serves a lot of people when compared with a centralized service. Publishing the user profiles in disseminated and centralized systems and providing recommendations is not safe. Most people dislike that their private details are stored in unsafe nodes all over the network. Besides, a recommender process is completely based on personal data that contains lots of private information. The concluding computation depends on the user's insight. The unwillingness of the user to provide likes and dislikes makes recommendations less accurate. Then, securing the user's privacy is fascinating for the user, since it can increase the accuracy of the RS [21].

5. Literature review

In [2], the authors described an abstract interaction model of a traditional RS and possible privacy breaches between the user and an RS. Existing methods that can protect privacy by RS are categorized as follows:

5.1 **Data masking method**

The major determination is to protect the privacy of folks recorded in the CF system. Noises are introduced to the user's data before interacting with the server for making recommendations. The users perturb their information and issue them to others, and it is the key parameter of RSs; if it is small, it cannot guarantee the purpose, while larger diminishes accuracy. One of the standard methods used for data-making is confusion, where a subset of the user's ratings is substituted by arbitrary values. The writers confirmed the performance of this approach does not affect accuracy [19]. This method is suggestible for e-commerce dealers, to distribute data about their customers without profaning their privacy [22]. In Ref. [23] the authors projected an approach to conclude the user's gender based on his/her preferences. However, these approaches can only be applicable among established users, and accumulation of noisy data is irrelevant for binary data (e.g., liked or disliked, thumb up or thumb down). Also, these methods fail to reveal relationships and degree of noise in the data.

5.2 **Differential privacy**

A mathematical and demonstrable privacy method assures for every record in the database, and the only access is via database queries. In this method, the recommendation task is performed by diminishing the loss between the rating matrix and noise to maintain adequate accuracy. In Ref. [24] the authors used the Laplace transformation approach for adding noise to the covariance matrix to safeguard the adjacent neighbors. In another approach, a relation-based graph recommendation was proposed and standardized the adjustment between accuracy and privacy [25].

5.3 **Secure multiparty computation**

This is reflected to be a subfield of cryptographic methods where multiple parties cooperatively agree on a function over their input while keeping them private. These approaches were familiarized to allow multiple parties or data distributors to mine their data together without revealing their data to each other. However, these methods fail to be used with large databases.

5.4 **Homomorphic encryption**

This approach can be divided into two types: IBCF and UBCF. In Ref. [26], the authors projected two privacy-preserving RSs for a mobile application built on trust assessment. Using the two schemes, they have preserved the users' privacy without losing the existing accuracy. In Ref. [6], the authors emphasized the challenges faced by differential privacy approaches. Nevertheless, in this method, the major weakness is that every user acquires the secret texts of another user's preferences through the recommendation process, and the privacy break escalates if the user conspires with the decrypted texts to learn further information. Author [27], proposed data packing to proficiently implement the cryptographic tasks on secret data by which computational and communication difficulty is reduced. However, this approach also grieves from the trade-off between accuracy and privacy issues. In Ref. [28], the authors addressed the method of multiplying two private data sets to create a recommendation for UBCF. However, in this approach the users have to produce extra private data, which increases the complexity.

According to the network structure of the system, we categorize RSs into (i) centralized and (ii) distributed. In real-world distribution, most current RSs are centralized in the logic that a service provider will compute the recommendations by collecting the preferences from all the users. The users' input data vary from explicit data such as *User X Item* matrix to implicit data such as the users' view list in the past or binary data such as thumb up/thumb down. This makes RS very secure intrusive to individual users. RS is the only entity that concerns preferable recommendations.

6. Quality measures for evaluating recommender systems

This is used when estimating an RS that is built on guessing the user's favorite against a dataset that articulates the user's approval in terms of ratings (such as Uber, Movierulz or Jokesbar) and generates predictions. Computing the exactness of predicted ratings corresponding to the provided ratings is a typical step for assessing recommender output.

In an information retrieval system, the occurrences are articles, and the responsibility is to suggest a set of related articles for the specified search term, or equivalently, to allocate every article to one of the two classes, "relevant" and "not relevant" In this instance, the "relevant" articles are merely those that fit into the "relevant" class. Recall and precision are measures that are used to measure the accuracy of the RS approach. Recall measure is described as the overall significant articles send back by a search divided by the overall current significant articles, whereas the measure precision is described as the overall significant articles send back by a search divided by the sum of articles returned by that search, as denoted following [6].

$$\text{Precision} = \frac{|TP|}{|TP| + |FP|}$$
$$\text{Recall} = \frac{|TP|}{|TP| + |FN|}$$

(7.5)

The matched items are considered true positive, and the differed items may belong to either false negative or false positive. The differed items from test data are false negative, and the differed items from training data are false positive. The relation between true positive, false positive, true negative, and false negative is mapped in a tabular representation known as a confusion matrix.

The mean absolute error (MAE) is the most widely used method of quality metric, alternatively called an absolute deviation. This approach solely considers the mean of the total dissimilarity between every rating and prediction of the users in the test set. MAE is useful for understanding the outcomes in a particular framework, on a scale of 1 to 5.

The root mean square error (RMSE) had been applied as a typical statistical metric to measure model performance in atmospheric sciences, air quality, and weather research studies. RMSE is an associated metric that has the consequence of assigning superior prominence on great deviations on a rating scale of 1 to 5. It is computed similarly as MAE, but it squares the error before summation, as shown following [6].

$$\text{MAE} = \frac{\sum\limits_{r_{ui} \neq .} |p_{ui} - r_{ui}|}{|\{r_{ui}/r_{ui} \neq .|}$$

$$\text{RMSE} = \sqrt{\frac{\sum\limits_{r_{ui} \neq .} (p_{ui} - r_{ui})^2}{|\{r_{ui}/r_{ui} \neq .\}|}}$$

(7.6)

7. Challenges in implementing recommender systems

Scalability: The computing speed rises as the amount of work by adding the users to the RS increases. The RS process works well when the number of the users is less, but scalability becomes tricky when this number has grown by more than a million users. This proportion of progress displays a linear relation and makes RSs challenging to process in wide-ranging data. The problem can be solved by making use of CF similar clustering feature and by decreasing dimensionality through deep learning.

Sparsity: An RS has numerous active users and products, nevertheless the user commonly prefers merely a small fraction of the accessible products. For example, MovieLens carries several groups of movies and requests new users to rate how much they like watching. However, the average user prefers just a couple of movies. The sparse entries in *User X Item* matrix generate computational difficulty and it is tough to compute neighboring users. To survive with this position, numerous methods can be used containing demographic cleaning and context-based CF algorithms.

Cold start problem: If the new the user/new item is included in the system, it will be difficult to recommend that product because of the lack of past history of that item. In such circumstances, neither the perception of the new user nor the new item can be expected, which leads to inaccurate recommendations. This problem can be resolved in several methods comprising (a) requesting the user to score some items initially, (b) requesting users to give their perception in aggregate, and (c) finally proposing favorite items.

Privacy: Providing private or secret information to the RS results in improved recommendation amenities but could lead to disputes of information security and privacy. Users are unwilling to provide their data into an RS that suffers from attack issues. Hence, an RS, either CF or CB, must build trusted paths, though the CF method is further likely to have such security concerns. In CF approaches, user ranking data are kept in a unified datastore, which may be negotiated causing data exploitation. For this motive, data encryption approaches can be used by supporting personalized recommendations without concerning mediators and associate users. Other practices include multidimensional trust model and multiprobing locality sensitive technique.

8. Basic scheme

In this scheme, m users, and n items are involved in rating to predict the target item. Encryption methods allow tasks to be done on ciphertext and produce encrypted effects. Formally, any encryption method $\in$ is valid for routes in $R_\in$ if for any key pair (PK, SK) the output by $GenKey_\in(\gamma)$ for any route $R \in R_\in$ any plaintext $p_1, p_2, \ldots, p_n$, and any ciphertext with $c_1, c_2, \ldots, c_n$ $C_i \leftarrow E(PK, p_i)$, e.g., $C \leftarrow Evaluate(PK, p_i, C_i)$ then $D(SK, c) \rightarrow C(p_1, p_2, \ldots, p_n)$.

Privacy preserving can be represented as a difficulty that data possessors wish to procure services without disclosing their private data. With this actuality, the chapter categorized the solutions in the following categories, as shown in Fig. 7.5.

Centralized scheme: In this case, a reliable server, which handles all the user data, desires to furnish recommendations to the users in its database, while securing the users' data from malicious attacks. These attacks allude to an attack method that illegally obtains knowledge about a tuple or database by examining statistical outputs [29].

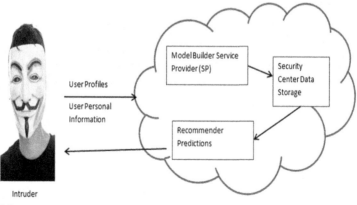

Intruder

FIGURE 7.5

System structure of centralized and distributed scheme.

Distributed scheme: In this case, numerous data possessors collaboratively accomplish a recommendation task, but none of them is willing to divulge their private data. Normally, the securing solutions are formulated without assuming a reliable server [29].

8.1 Centralized scheme

In a centralized scheme, the users must register at a reliable server and publish their ratings to the service provider (SP). Next, the SP computes transitional data for a seeker, and the recommender encodes this information with the privacy-preserving technique and sends this encrypted data to the SP [26]. The SP computes decryption calculations and generates final recommendations. This scheme includes three phases: system initialization, database modeling, and cooperative recommendation generation [2,30]. Exhaustive details of each phase are described in Fig. 7.6.

System bootup: It is a process of user enrollment and has the subsequent steps: Initially, the users register at the reliable server by directing $sig_i(uid_i)$ a unique signature to SC, where $sig_i(uid_i)$ is the user's identity [31]. Secondly, SC acknowledges the user's enrollment by returning back its signature on the user's identity (i.e., $sig_{SC}(sig_i(uid_i))$). Besides, SC generates a public key (i.e., pk) and an equivalent private key (i.e., sk) (i.e., $(pk, sk) \leftarrow G(SC)$), and then SC chooses a random number r protected by encryption and composed with $sig_{SC}(sig_i(uid_i))$.

Database modeling: The users' private data are gathered and $EK(r_1)$ is calculated by showing $r_1 = H(r)$ with $EK(r_1) = EK\{pk_{SC}, r_1\}$, where pk_{SC} is the public key of SC [32]. Formerly, this value was multiplied by $EK(r_1)$ and the user's identity at a time t. This generated data is stored in SP for further processing.

In the next time slice, the database will be reorganized as, initially, the user's calculated $r_{t+1} = H(r_t)$, and multiplied by with the user's identity.

Cooperative recommendation generation. This is a collaborating process comprising the recommender campaigner, SC, and SP. Recommender campaigner directs an appeal to SP initially, and a set of correlated user lists are directed back

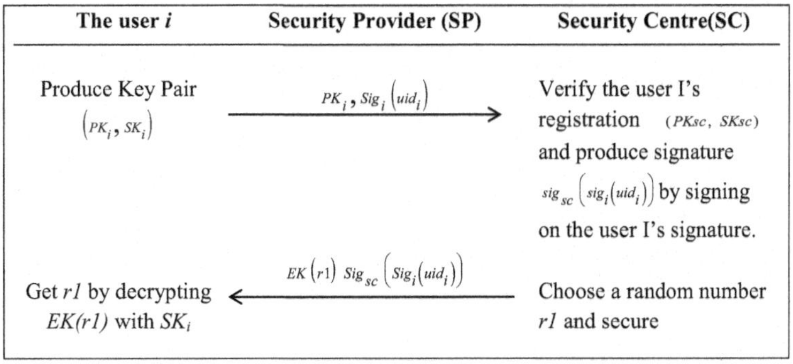

The user i	Security Provider (SP)	Security Centre(SC)
Produce Key Pair $\left(PK_i, SK_i\right)$	$PK_i, Sig_i\left(uid_i\right)$ $\longrightarrow$	Verify the user I's registration (PK_{sc}, SK_{sc}) and produce signature $sig_{sc}\left(sig_i\left(uid_i\right)\right)$ by signing on the user I's signature.
Get $r1$ by decrypting $EK(r1)$ with SK_i	$\longleftarrow$ $EK\left(r1\right) Sig_{sc}\left(Sig_i\left(uid_i\right)\right)$	Choose a random number $r1$ and secure

FIGURE 7.6

Basic initialization of centralized scheme.

if the user's identity validation is accurate. After that, the recommender campaigner computes the user associations with the homomorphic operation and sends the outcome to SP. Finally, SP decrypts the outcome using homomorphic operation for producing final recommendations.

8.2 Distributed scheme

This scheme is acceptable for cloud service surroundings. This scheme too includes three phases: system bootup, database modeling, and cooperative recommendation production. Exhaustive details of each phase are described subsequently. Nevertheless, the process in each phase is different from the centralized scheme [2,30]. All the users appear as a group in system bootup, and a most trusted the user is elected to generate a secret key and distribute among others based on security algorithm. This method produces a valid number of subsecret keys among the group. Database modeling is activated when a trusted user directs a recommendation campaigner to SP, and the user's data secured under this secret is stored in SP for producing recommendations, as shown in Fig. 7.7. A cooperative recommendation is managed between recommendation campaigner and SP to produce concluding recommendations [26].

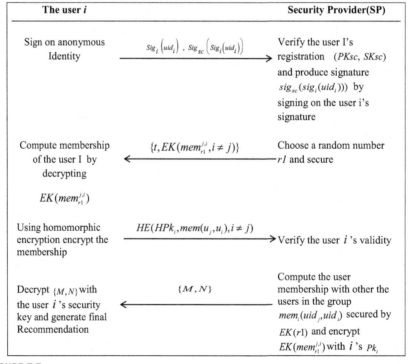

FIGURE 7.7

Basic initialization of distributed scheme.

System bootup: After the users' registration, the users form a group as maintained by common properties or social relationships, which is common in real life [5]. Another instance is a public cluster that might be created implicitly when the users follow a usual post in social sites or have the same interests such as religion, customs, or norms to obtain the recommendations from other the users.

Database modeling: This phase starts when the user raises a recommendation appeal to the SP and takes a definite period according to the user registration and system strategy. Consider the user i for an instance, and the process of database modeling is defined as the following: Target user i should be a candidate of the group and notifies his private key to all the members of the group. Group users receive his private key and encrypt r into $EK(r)$ with the user i's public key and store these details in the database. After validating the user i's registration at a reliable server the recommendation request is forwarded to the SP. Finally, the SP sends this result to the entire group. Then, users multiply this with $EK(r)$ and forward with their identities to the SP [2,30].

Cooperative recommendation generation: This process is identical to the centrality scheme, except that the user's validity is replaced by the group's validity. By applying encryption and decryption on the user's data, one can keep the SP from guessing final recommendations for a target user.

8.3 Framework of securing rating matrix, similarity, and aggregate computations in centralized and distributed schemes

A rating matrix is the basic input for any RS; it is a measure of the users' preference of an item on a sliding scale between 1 and 5. If the system explores this matrix, intruders can gain access and attempt to evaluate like-minded users, favoring his/her recommendation by an increase or decrease in the rating value and suggesting erroneous recommendations. To overcome this difficulty, several approaches were proposed.

One such approach is homomorphic encryption, which is a popular semantically secure privacy-preserving method and computationally very hard to determine plaintext from a ciphertext for any intruder. Additionally, this method ensures security without losing accuracy.

Homomorphic addition: Homomorphic addition is the ciphertext of the summation of their equivalent plaintexts [30,33].

$$C_1 * C_2 = (pk^{m_1} sk^{r_1}) * (pk^{m_2} sk^{r_2}) = pk^{m_1+m_2} sk^{r_1+r_2}, C_1 = HE(m_1) \text{ and } C_2$$
$$= HE(m_2) \tag{7.7}$$

$$HE(C_1, C_2) = m_1 + m_2$$

Homomorphic multiplication: Multiplication of two ciphertexts $c1$ and $c2$ is also possible and is defined next [33]:

$$HD((C_1)^c) = HD(pk^m)^c = m * c \tag{7.8}$$

8.3.1 Collaborative recommender model

To depict this considers the following rating matrix with n users and m items in an RS. The user–item rating matrix is signified by UI_M in which r_{ij} denotes the rating produced by the user i for a product j. To secure this matrix the system generates the user's group, a set of prime numbers with key generators g. The public and private keys are signified by $\{PK, SK\}$, and the secret key is denoted by $SK = pk^m$, where m has integer numbers in $\{0, 1, ...n\}$.

For n the users, the system computes a combined public key (i.e., $PK = \{pk_1, pk_2, ..., pk_n\}$) and explores the remaining users to encrypt their ratings.

$$PK = pk_1, pk_2, ..., pk_n = \prod_{i=1}^{n} pk_i = pk^{\sum_{i=1}^{n} m_i} \tag{7.9}$$

To encrypt the message m, a random number r_1 and a public key are chosen and the ciphertext is calculate as $C \leftarrow pk^m sk^{r_1}$. The decryption is computed for this encrypted message as $D \leftarrow (pk^m sk^{r_1})^{q_i} \leftarrow (pk^{q_1})^m$. The detailed steps to encrypt the rating matrix are outlined next:

Step 1: Each user encrypts their item ratings as $pk^{r_{ij}}$ and $pk^{t_{ij}}$ respectively. They create an integrated message $M_{(1)}$ comprising encrypted ratings and turns, which are dispatched to the server, where $r_{i,j}$ signifies a rating given by the user u_i for item i_i and sets its value to either 0 or 1 based on the availability of the rating, as shown in Table 7.4. $M_{(1)} = \{A_{1j}, A_{2j}, A_{3j}\}$ where $j = 1, 2....m$.

$$M_{(1)} = \left\{ \begin{array}{c} EK(pk^3), EK(pk^1) \\ EK(pk^2), EK(pk^1) \\ EK(pk^0), EK(pk^0) \end{array} \right\}$$

$$M_{(2)} = \left\{ \begin{array}{c} EK(pk^5), EK(pk^1) \\ EK(pk^1), EK(pk^1) \\ EK(pk^3), EK(pk^1) \end{array} \right\} \tag{7.10}$$

$$M_{(3)} = \left\{ \begin{array}{c} EK(pk^5), EK(pk^1) \\ EK(pk^0), EK(pk^0) \\ EK(pk^2), EK(pk^1) \end{array} \right\}$$

Table 7.4 Rating matrix.

	i_1	i_2	i_3	i_4
u_1	3	5	5	4
u_2	2	1	0	4
u_3	0	3	2	0

Step 2: The server calculates the homomorphic addition of the ratings and turns as follows:

$$\left(A_{1,j}, B_{1,j}\right) = \prod_{k=1}^{n} EK(pk^{r_{i,j}}) \tag{7.11}$$

$$\left(A_{2,j}, B_{2,j}\right) = \prod_{j=1}^{n} EK(pk^{t_{i,j}})$$

Where $\left(A_{1,j}, B_{1,j}\right)$ and $\left(A_{2,j}, B_{2,j}\right)$ denote the ciphertexts and the server transmits this to other users by $M_{(2)} = \{A_{1,j}, A_{2,j}\}$.

Step 3: All the users decrypt the message by using their secret keys and transmit the following message back to the server.

$$M_{(3)} = \left\{ \left(A_{1,j}\right)^{qi}, \left(A_{2,j}\right)^{qi} \right\} \text{ where } j = 1, 2....m \tag{7.12}$$

Step 4: The system decrypts the encrypted message as described next:

$$C_{1,j} = \frac{B_{1,j}}{\prod\limits_{i=1}^{n} \left(A_{1,j}\right)^{qi}} C_{2,j} = \frac{B_{2,j}}{\prod\limits_{i=1}^{n} \left(A_{2,j}\right)^{qi}} \tag{7.13}$$

The system discovers the plain messages by calculating using

$$D_{1,j} = \log_{pk} C_{1,j} \text{ and } D_{2,j} = \log_{pk} C_{2,j} \text{ respectively.} \tag{7.14}$$

From the preceding rating matrix, the users' messages, as in Eq. (7.10), are forwarded to the system. Once the system obtains ratings and turns messages, it calculates the total ratings acquired for items as follows.

$$M_{(1)} = \left(A_{1,1}, B_{1,1}\right) = EK\left(pk^{3}\right).EK\left(pk^{2}\right).EK\left(pk^{0}\right) = EK\left(pk^{5}\right) \tag{7.15}$$

$$M_{(1)} = \left(A_{2,1}, B_{2,1}\right) = EK\left(pk^{1}\right).EK\left(pk^{1}\right).EK\left(pk^{0}\right) = EK\left(pk^{2}\right)$$

The users send this message $M_{(1)} = \left\{ \left(A_{1,1}\right)^{qi}, \left(A_{2,1}\right)^{qi} \right\}_{\{i=1,2,3...n\}}$ to the server, and the server gets the plaintext of ratings and turns as follows:

$$C_{1,1} = \frac{B_{1,1}}{\prod\limits_{i=1}^{3} \left(A_{1,1}\right)^{qi}} = pk^{5} \quad \text{and} \quad C_{2,1} = \frac{B_{2,1}}{\prod\limits_{i=1}^{3} \left(A_{2,1}\right)^{qi}} = pk^{2} \tag{7.16}$$

$D_{1,1} = \log_{pk} C_{1,1} = \log_{pk} pk^{5} = 5$ and $D_{1,2} = \log_{pk} C_{2,1} = \log_{pk} pk^{2} = 2$

A recommender model that captures the users' preferences relies on a metric function that computes similarity among the users and items. One metric to compute the similarity of two items is the cosine metric of Eq. (7.17):

$$sim(\bar{i}, \bar{j}) = \frac{\bar{i}.\bar{j}}{|\bar{i}||\bar{j}|} = \frac{\sum i_k j_k}{\sqrt{\sum i_k^2 \sum j_k^2}} \tag{7.17}$$

8.3.1.1 Securing item–item similarity computation

In this illustration, we demonstrate how to compute the correspondence between items i_2 and i_3 protectively. All the users direct encrypted messages to the system [32]. For example, the users send the following ciphertexts for the item i_2 and i_3 to the system as follows:

$$M^1_{(2,3)} = \left\{ EK\left(pk^{5.0}\right), EK\left(pk^{5^2}\right), EK\left(pk^{0^2}\right) \right\}$$

$$M^2_{(2,3)} = \left\{ EK\left(pk^{1.5}\right), EK\left(pk^{1^2}\right), EK\left(pk^{5^2}\right) \right\} \tag{7.18}$$

$$M^3_{(2,3)} = \left\{ EK\left(pk^{3.2}\right), EK\left(pk^{3^2}\right), EK\left(pk^{2^2}\right) \right\}$$

At the other end, after receiving these ciphertexts by the system, it calculates homomorphic ciphertexts between the items as follows:

$$\left(A_{2,3}, B_{2,3}\right) = EK\left(pk^{5.0}\right).EK\left(pk^{1.5}\right).EK\left(pk^{3.2}\right) = EK\left(pk^{11}\right)$$

$$\left(A_2, B_1\right) == EK\left(pk^{5^2}\right).EK\left(pk^{1^2}\right).EK\left(pk^{3^2}\right) = EK\left(pk^{35}\right) \tag{7.19}$$

$$\left(A_2, B_2\right) == EK\left(pk^{0^2}\right).EK\left(pk^{5^2}\right).EK\left(pk^{2^2}\right) = EK\left(pk^{29}\right)$$

The system calculates all item pairs similarities and distributes a share of these encrypted texts to every the user as

$$M_{(2,3)} = \left\{ A_{2,3}, A_2, A_3 \right\} \tag{7.20}$$

Collaboratively all the users decrypt the ciphertexts and send the message to the system by

$$M_{(2,3)} = \left\{ \left(A_{2,3}\right)^{q_i}, \left(A_2\right)^{q_i}, \left(A_3\right)^{q_i} \right\} \tag{7.21}$$

Where q_i denotes the secret key of every user. The system decrypts this similarity message between i_2 and i_3 as

$$C_{2,1,3} = \frac{B_{2,3}}{\left(\prod_{i=1}^{3} A_{2,3}\right)^{q_i}} = pk^{11}$$

$$C_{2,1} = \frac{B_2}{\left(\prod_{i=1}^{3} A_2\right)^{q_i}} = pk^{35} \tag{7.22}$$

$$C_{3,1} = \frac{B_3}{\left(\prod_{i=1}^{3} A_3\right)^{q_i}} = pk^{29}$$

The resultant similarities are derived as $D_{2,3} = \log_{pk} pk^{11} = 11$, $D_2 = \log_{pk} pk^{35} = 35$, and $D_3 = \log_{pk} pk^{29} = 29$. The cosine similarity between items is calculated by the system using Eq. (7.17) as follows:

$$s(i_2, i_3) = \frac{11}{\sqrt{35}\sqrt{29}} = 0.346$$

Item–item similarities are computed between all the other items, and the subsequent matrix is kept by the system, as shown in Table 7.5. This similarity range between any two items resides between -1 and 1, as shown in Table 7.5.

8.3.1.2 Generating recommendations

For example, consider u_1 as a target user on item i_1, and the system generates recommendations as follows. Target user u_1 directs his encrypted ratings, which are multiplied by 100 to manage the ElGamal cryptosystem.

$$M_{(1)} = \left\{ EK(pk^{300}), EK(pk^{500}), EK(pk^{500}), EK(pk^{400}) \right\} \tag{7.23}$$

The system calculates numerator and denominator for cosine similarity metric as

$$EK\left(pk^{250(89+19+97)}\right) \text{ and}$$

$$\left(EK(pk^{500})/EK(pk^{300})\right)^{89} \cdot \left(EK(pk^{500})/EK(pk^{350})\right)^{19} \cdot \left(EK(pk^{400})/EK(pk^{400})\right)^{97} \tag{7.24}$$

The system finally calculates homomorphic encryption as shown:

$$(A_1, B_1)^{(1)} = EK\left(pk^{250(89+19+97)}\right) \cdot \left(EK(pk^{500})/EK(pk^{300})\right)^{89}$$
$$\cdot \left(EK(pk^{500})/EK(pk^{350})\right)^{19} \cdot \left(EK(pk^{400})/EK(pk^{400})\right)^{97}$$
$$= EK(pk^{63050}) \tag{7.25}$$

$$(A_1, B_1)^{(2)} = EK(pk^{205})$$

Every user in the group obtains these homomorphic ciphertexts and decrypts them using their secret keys by

$$C_{(1,1)} = EK(pk^{63050}) \, C_{(1,2)} = EK(pk^{205})$$

Table 7.5 Item–item similarity matrix.

Item	i_1	i_2	i_3	i_4
i_1	100	89	19	97
i_2		100	34	94
i_3			100	25
i_4				100

$$D_{(1,1)} = \log_{pk} pk^{63050} = 63050 D_{(1,2)} = \log_{pk} pk^{205} = 205 \qquad (7.26)$$

The final prediction is calculated as

$$P_{(1,1)} = 307.56 = 3.07$$

Likewise, we obtain the remaining item predictions for u_1 as $P_{(1,2)} = 2.56$, $P_{(1,3)} = 4.34$, and $P_{(1,4)} = 4.56$. The generated predictions build a matrix, and Chinese remainder theorem encryption and decryption can be performed to protect the highest predictions. As item i_4 has a great prediction value, it is recommended for the target user.

8.3.1.3 Securing the user–user similarity computation

We demonstrate how to calculate the similarity between the users u_1 and u_2 protectively. All the users direct encrypted messages to the system [33]. For example, the users send the following ciphertexts for the users u_1 and u_2 to the system as follows:

$$
\begin{aligned}
M^{(1)} &= \left(EK(pk^{30}), EK\left(pk^{3^2}\right), EK\left(pk^{0^2}\right)\right\} \\
M^{(2)} &= \left(EK(pk^{51}), EK\left(pk^{5^2}\right), EK\left(pk^{1^2}\right)\right\} \\
M^{(3)} &= \left(EK(pk^{05}), EK\left(pk^{0^2}\right), EK\left(pk^{5^2}\right)\right\} \\
M^{(4)} &= \left(EK(pk^{40}), EK\left(pk^{4^2}\right), EK\left(pk^{0^2}\right)\right\}
\end{aligned}
\qquad (7.27)
$$

At the other end, after receiving these ciphertexts by the system, it calculates homomorphic ciphertexts between the users as follows:

$$
\begin{aligned}
(A_{12}, B_{12}) &= EK(pk^{3.0}).EK(pk^{5.1}).EK(pk^{0.5}).EK(pk^{4.0}) = EK(pk^5) \\
(A_1, B_1) &= EK\left(pk^{3^2}\right).EK\left(pk^{5^2}\right).EK\left(pk^{0^2}\right).EK\left(pk^{4^2}\right) = EK(pk^{50}) \\
(A_2, B_2) &= EK\left(pk^{0^2}\right).EK\left(pk^{1^2}\right).EK\left(pk^{5^2}\right).EK\left(pk^{0^2}\right) = EK(pk^{26})
\end{aligned}
\qquad (7.28)
$$

The system calculates all the user–user pair similarities and distributes a share of these ciphertexts to all the users as

$$M_{(1,2)} = \left\{A_{1,2}, A_1, A_2\right\} \qquad (7.29)$$

Collaboratively, all the users decrypt the encrypted text and deliver the same to the system by

$$M_{(1,2)} = \left\{(A_{1,2})^{q_i}, (A_1)^{q_i}, (A_2)^{q_i}\right\} \qquad (7.30)$$

Where q_i denotes the secret key of every user. The system decrypts this similarity message between u_1 and u_2 as follows:

$$C_{2,1} = \frac{B_{1,2}}{\left(\prod_{i=1}^{3} A_{1,2}\right)^{q_i}} = pk^5$$

$$C_{1,2} = \frac{B_1}{\left(\prod_{i=1}^{3} A_1\right)^{q_i}} = pk^{50}$$

$$C_{2,1} = \frac{B_2}{\left(\prod_{i=1}^{3} A_2\right)^{q_i}} = pk^{26}$$

(7.31)

The resultant similarities are derived as $D_{1,2} = \log_{pk} pk^5 = 5$, $D_1 = \log_{pk} pk^{50} = 50$, and $D_2 = \log_{pk} pk^{26} = 26$. The cosine similarity between the users u_1 and u_2 is calculated. Likewise the user–user similarities are computed by the system as follows:

$$s(i_1, i_2) = \frac{5}{\sqrt{50}\sqrt{26}} = 0.138$$

8.3.2 Content-based recommender model

To depict this model considers the following matrix with m articles and n words in an RS [34], as shown in Table 7.6. The article–word frequency matrix is represented by AW_M in which $f_{i,j}$ represents the occurrence of a word j in the article i. The detailed steps to encrypt the article–word matrix are outlined in Table 7.6.

Step 1: The user sends the encrypted data to the system as follows:

$$M_{(1)} = \left\{ \begin{array}{l} EK\left(pk^{21}\right), EK\left(pk^1\right) \\ EK\left(pk^{24}\right), EK\left(pk^1\right) \\ EK\left(pk^0\right), EK\left(pk^0\right) \\ EK\left(pk^2\right), EK\left(pk^1\right) \end{array} \right\}$$

$$M_{(2)} = \left\{ \begin{array}{l} EK\left(pk^{24}\right), EK\left(pk^1\right) \\ EK\left(pk^{59}\right), EK\left(pk^1\right) \\ EK\left(pk^2\right), EK\left(pk^1\right) \\ EK\left(pk^1\right), EK\left(pk^1\right) \end{array} \right\}$$

(7.32)

$$M_{(3)} = \left\{ \begin{array}{l} EK\left(pk^{40}\right), EK\left(pk^1\right) \\ EK\left(pk^{115}\right), EK\left(pk^1\right) \\ EK\left(pk^8\right), EK\left(pk^0\right) \\ EK\left(pk^{10}\right), EK\left(pk^1\right) \end{array} \right\}$$

Table 7.6 Article–word frequency matrix.

	w_1	w_2	w_3	w_4
a_1	21	24	0	2
a_2	24	59	2	1
a_3	40	115	8	10
a_4	4	28	5	0
.....				
DF	5000	50,000	10,000	7000

The frequency count corresponding to every word for each article is computed by the system (document frequency, i.e., DF).

$$(A_{1,1}, B_{1,1}) = EK(pk^{21}).EK(pk^{24}).EK(pk^0).EK(pk^2)... = EK(pk^{5000}) \quad (7.33)$$

$$(A_{1,1}, B_{1,1}) = EK(pk^1).EK(pk^1).EK(pk^0).EK(pk^1)... = EK(pk^{61})$$

The server transmits this encrypted message for all articles to all words by $M_{(1)} = \{A_{1,1}, A_{2,1}\}$.

Step 2: All the users decrypt the message by using their secret keys and transmit the resulting message back to the server.

$$M_{(1)} = \left\{ (A_{1,1})^{qi}, (A_{2,1})^{qi} \right\} \text{ where } i = 1, 2....m \quad (7.34)$$

Step 3: The system decrypts the ciphertexts described next:

$$C_{1,1} = \frac{B_{1,1}}{\prod_{i=1}^{n}(A_{1,1})^{qi}} = pk^{5000} C_{1,2} = \frac{B_{2,2}}{\prod_{i=1}^{n}(A_{2,1})^{qi}} = pk^{61} \quad (7.35)$$

The system decrypts the plain messages and document frequency by calculating using

$D_{1,1} = \log_{pk} C_{1,1}$, $D_{1,2} = \log_{pk} C_{1,2}$ and $D_{(1,1)}^{(1)} = \log_{pk} pk^{21} = 21$, $D_{(1,2)}^{(1)} = \log_{pk} pk^{24} = 24$ respectively.

$$D_{1,1} = \log_{pk} C_{1,1} = \log_g pk^{5000} = 5000, D_{1,2} = \log_{pk} C_{2,1} = \log_g pk^{11} = 61 \quad (7.36)$$

$$D_{2,1} = 50000, D_{3,1} = 10000$$

The average of each word is computed for all other words as

$$avg(R_1) = \frac{E(pk^{5000})}{E(pk^{11})}$$

Step 4: The word frequency for all the other words of each of the articles is computed and distributed as shown:

$$D_{(1,1)}^{(2)} = 1 + \log_g \left(D_{(1,1)}^{(1)} \right) = 2.32$$

$$D_{(1,2)}^{(2)} = 1 + \log_g \left(D_{(2,1)}^{(2)} \right) = 2.37 \tag{7.37}$$

$$D_{(1,3)}^{(2)} = 1 + \log_g \left(D_{(3,1)}^{(3)} \right) = 2.60$$

The final computation is shown in Table 7.7, if we assume the generator as 10.

Next, IDF is computed (as in Eq. 7.38) by taking the inverse of the DF among the entire corpus as shown in Table 7.5. So, if the corpus size is one lakh, the IDF score for each word will be computed as shown [31].

$$(A_{I1}, B_{I1}) = \left(EK \left(pk^{50000} \right).EK \left(pk^{10000} \right).EK \left(pk^{7000} \right) \right)/EK \left(pk^{5000} \right) \tag{7.38}$$

$$D_{(1,1)}^{(3)} = \left(\log_g pk^{20} \right) = 1.30$$

The IDF process assigns a lower value for the most frequent word. The length of this vector is equivalent to the square of the sum of the squared weights of all words, which are computed as

$$(A_{ID1}, B_{ID1}) = EK \left(pk^{232^2} \right).EK \left(pk^{237^2} \right).EK \left(pk^{0^2} \right).EK \left(pk^{131^2} \right)$$

$$= pk^{\sqrt{M_{(1)}^{(2)}}} = 3.80 \tag{7.39}$$

Finally, the term vector is divided by the length of the vector to get a regularized vector. So, for article 1, the regularized vector score is

$$(A_{T1}, B_{T1}) = EK \left(pk^{232} \right)/EK \left(pk^{380} \right) = pk^{6105} \tag{7.40}$$

$$D_{(1,1)}^{(4)} = \log_{pk} pk^{6105} = 6105$$

Finally, the similarity between articles is computed using cosine similarity as follows [34]:

$$(A_{12}, B_{12}) = EK \left(pk^{610.594} \right).EK \left(pk^{626.691} \right).EK \left(pk^{0.342} \right) = EK \left(pk^{797} \right) \tag{7.41}$$

Table 7.7 Word frequencies of all words.

	w_1	w_2	w_3	w_4
a_1	2.32	2.37	0	1.31
a_2	2.37	2.75	1.31	1
a_3	2.61	3.06	1.91	2
a_4	1.61	2.45	1.68	0
.....				

At the following phase, the user's neighborhood cluster comprises the maximum number of similar adjacent users, which is produced with the seeker [5,13] Lastly, a prediction is produced based on the items' aggregated average of this cluster, which results in a reference list of the top predicted ratings.

8.3.3 Trust-aware recommendations

Traditional RS includes users with untrusted declarations, for which the predictions of anonymous preferences cannot be guessed [35]. For instance, if confidential relations do not matter to the user, it is difficult to guess unknown items' ratings. Thus, by expending the user's trust, one can improve the coverage [35].

The trust can be depicted as a weighted directed network for which nodes signify the users and an edge among the users represents the accessibility of their trusted relations. The trust between any two users is computed as follows:

$$t_{u,v} = \left(\frac{d_m - d_{u,v} + 1}{d_m} \right) \tag{7.42}$$

Where d_m is the propagation distance, and $d_{u,v}$ is the distance between the users.

Trusted rating imputation:

Unknown rating prediction is the solution for data sparsity that persuades improving the performance of RS. For highly sparse matrices, comprising trust propagation is a reliable solution and is processed as shown [36]:

$$r_{u,i} = \frac{\sum_{TF} t_{u,v} * r_{v,i}}{\sum_{TF} t_{u,v}} \tag{7.43}$$

where $r_{u,i}$ is the guessed rating, and TF signifies user u's trusted users. A reliability metric is required to evaluate the worth of the predicted rating, which is defined as

$$C_{u,i} = \frac{1}{2} \int_0^1 \left| \frac{x^{positive}(1-x)^{negitive}}{\int_0^1 x^{positive}(1-x)^{negitive} dx} - 1 \right| dx \tag{7.44}$$

where $C_{u,i}$ is the reliability value that resides in the range between [0,1], $x^{positive}$ signifies the users with positive feedback over items, and the users with negative feedback are denoted with $(1-x)^{negitive}$. It is observed that positive feedback is higher than 3 and negative feedback is less than 3. Classical Pearson correlation is slightly modified to incorporate reliability in the similarity between the users as follows [37].

$$UC_{u,v} = \frac{\sum_{i \in I_{u,v}} C_{u,i}(r_{u,i} - \bar{r}_u) C_{v,i}(r_{v,i} - \bar{r}_v)}{\sqrt{\sum_{i \in I_{u,v}} C_{u,i}^2(r_{u,i} - \bar{r}_u)^2} \sqrt{\sum_{i \in I_{u,v}} C_{v,i}^2(r_{v,i} - \bar{r}_v)^2}} \tag{7.45}$$

Table 7.8 Computation score of IDF.

DF	5000	50,000	10,000	7000
IDF	1.30	0.30	1	1.14
N	100,000			

Table 7.9 Regularized vector score between all articles.

	w_1	w_2	w_3
a_1	0.6105	0.626	0
a_2	0.594	0.691	0.342
a_3	0.481	0.568	0.353

Table 7.10 Similarity score between articles.

	Cosine similarity
$\cos(a_1, a_2)$	0.797
$\cos(a_2, a_3)$	0.794
$\cos(a_1, a_3)$	0.65

where $UC_{u,v}$ is the metric to identify the commonly preferred items by both users. Next, Pareto dominance concepts are used to find principal users, which are erased from the prediction procedure of a target user [38]. Only nondominant users are considered neighbors while computing the final trust as follows [36]:

$$TW_{u,v} = t_{u,v} * UC_{u,v} \tag{7.46}$$

where $TW_{u,v}$ is the final trust value among the users u and v. Finally, for target users a top list of recommendations are predicted based on the following:

$$pr_{u,i} = \frac{\sum_{TND} TW_{u,v} * r_{v,i}}{\sum_{TND} TW_{u,v}} \tag{7.47}$$

where TND is a list of nondominant users.

8.3.4 Identifying attack paths

An RS structure naturally involves nodes that can be misused by intruders to gain entry into the network. Besides, the propagation distance and the number of exposures are the basic issues that regulate the scope of the attack graph [37]. When this scope increases the likelihood of more possibilities for an intruder also increases. This cyber-attack prevention method includes the following steps:

Step 1: Identify various activities (e.g., valuable assets, attacker profiles).

Step 2: For each valuable asset, identify the attacked path and affected assets.

2.1 If the attacker location and attacker capability is more than expected severity, identify the type of vulnerability [39].

Step 3: Identify the shortest path distance in a graph. Set propagation length to target asset position.

Step 4: Add identified attacked paths and affected assets to a list to prevent attacks within a risk management system.

9. Conclusion

This chapter presents privacy-preserving practices for item-based, user-based CF and content-based CF. In this proposed approach, you may perceive desirable features like the users' privacy and the generation of recommendations. By performing calculations on encrypted data, an RS secures all possible access paths concerning the system and the user. Conferring to the suggested approaches, neither the users nor the systems can study any user's profiles, ratings, or neighboring information comprising any approach to reveal intermediate recommendation outcomes. We show encrypted approaches to secure rating matrix, similarity, and aggregate computations, encrypting top items, building trusted networks without losing accuracy with regard to the recommendation.

References

[1] M. Morita, Y. Shinoda, Information filtering based on the user behavior analysis and best match text retrieval, in: Proceedings of the 17th Annual International ACM SIGIR Conference on Research and Development in Information Retrieval, Springer-Verlag New York, Inc, August 1994, pp. 272–281.

[2] C. Wang, Y. Zheng, J. Jiang, K. Ren, Toward privacy-preserving personalized recommendation services, Engineering 4 (1) (2018) 21–28, 15. Privacy Preserving The user-based RS.

[3] H.J. Ahn, A new similarity measure for collaborative filtering to alleviate the new the user cold-starting problem, Inf. Sci. 178 (1) (2008) 37–51.

[4] G. van Capelleveen, C. Amrit, D.M. Yazan, H. Zijm, The recommender canvas: a model for developing and documenting RS design, Expert Syst. Appl. 129 (2019) 97–117, 13. A Practical Privacy-Preserving RS.

[5] R. Katarya, O.P. Verma, Privacy-preserving and secure RS enhance with K-NN and social tagging, in: 2017 IEEE 4th International Conference on Cyber Security and Cloud Computing (CSCloud), IEEE, June 2017, pp. 52–57.

[6] J. Wang, Q. Tang, RS and Their Security Concerns, 2015.

[7] X. Su, T.M. Khoshgoftaar, A survey of collaborative filtering techniques, Adv. Artif. Intell. 2009 (2009) 4.

[8] T. Bogers, A. Van den Bosch, in: Collaborative and Content-Based Filtering for Item Recommendation on Social Bookmarking Websites. Submitted to CIKM, 2009, p. 9.

[9] P. Lops, M. De Gemmis, G. Semeraro, Content-based RS: state of the art and trends, in: RS Handbook, Springer, Boston, MA, 2011, pp. 73−105.

[10] H.C. Wu, R.W.P. Luk, K.F. Wong, K.L. Kwok, Interpreting tf-idf term weights as making relevance decisions, ACM Trans. Inf. Syst. 26 (3) (2008) 13.

[11] B.D. Deebak, F. Al-Turjman, A Novel Community-Based Trust Aware Recommender Systems for Big Data Cloud Service Networks, Sustain. Cities Soc. (2020) 102274.

[12] J. Kamahara, T. Asakawa, S. Shimojo, H. Miyahara, A community -based recommendation system to reveal unexpected interests, in: Multimedia Modelling Conference, 2005. MMM 2005. Proceedings of the 11th International, IEEE, January 2005, pp. 433−438.

[13] L. Guo, Q. Peng, A neighbor selection method based on network community detection for collaborative filtering, in: Computer and Information Science (ICIS), 2014 IEEE/ACIS 13th International Conference on, IEEE, June 2014, pp. 143−148.

[14] L. Gao, C. Li, Hybrid personalized recommended model based on genetic algorithm, in: 2008 4th International Conference on Wireless Communications, Networking and Mobile Computing, IEEE, October 2008, pp. 1−4.

[15] X. Shi, H. Ye, S. Gong, A personalized recommender integrating item-based and the user-based collaborative filtering, in: 2008 International Seminar on Business and Information Management, vol. 1, IEEE, December 2008, pp. 264−267.

[16] B. Sarwar, G. Karypis, J. Konstan, J. Riedl, Item-based collaborative filtering recommendation algorithms, in: Proceedings of the 10th International Conference on World Wide Web, ACM, April 2001, pp. 285−295.

[17] S. Ahmadian, M. Meghdadi, M. Afsharchi, A social recommendation method based on an adaptive neighbor selection mechanism, Inf. Process. Manag. 54 (4) (2018) 707−725.

[18] F. Ricci, L. Rokach, B. Shapira, Introduction to RS handbook, in: RS Handbook, Springer US, 2011, pp. 1−35.

[19] H. Polat, W. Du, Privacy-preserving collaborative filtering using randomized perturbation techniques, in: Third IEEE International Conference on Data Mining, 2003, pp. 625−628.

[20] V. Arnaboldi, M.G. Campana, F. Delmastro, E. Pagani, A personalized RS for pervasive social networks, Pervasive Mob. Comput. 36 (2017) 3−24.

[21] S. Jabbar, S. Khalid, M. Latif, F. Al-Turjman, L. Mostarda, Cyber Security Threats detection in Internet of Things using Deep Learning approach, IEEE Access 7 (1) (2019) 124379−124389.

[22] K. Wei, J. Huang, S. Fu, A survey of e-commerce RS.. 2007 International Conference on Service Systems and Service Management, IEEE, June 2007, pp. 1−5.

[23] U. Weinsberg, S. Bhagat, S. Ioannidis, N. Taft, BlurMe: inferring and obfuscating the user gender based on ratings, in: Proceedings of the Sixth ACM Conference on RS, ACM, September 2012, pp. 195−202.

[24] F. McSherry, I. Mironov, Differentially private RS: building privacy into the netflix prize contenders, in: Proceedings of the 15th ACM SIGKDD International Conference on Knowledge Discovery and Data Mining, ACM, June 2009, pp. 627−636.

[25] D. Deebak, F. Al-Turjman, L. Mostarda, Seamless Secure Anonymous Authentication for Cloud-Based Mobile Edge Computing, Elsevier Comput. Electr. Eng. J. (2020).

[26] K. Xu, W. Zhang, Z. Yan, A privacy-preserving mobile application RS based on trust evaluation, J. Comput. Sci. 26 (2018) 87−107.

[27] A. Murugesan, B. Saminathan, F. Al-Turjman, R. Lakshmana, Analysis On Homomorphic Technique For Data Security In Fog Computing, Wiley Transactions on Emerging Telecommun. Technol. (2020), https://doi.org/10.1002/ett.3990.

[28] A.T. Fadi, D.B. David, Seamless Authentication: For IoT-Big Data Technologies in Smart Industrial Application Systems, IEEE Trans. Ind. Inform. (2020).

[29] K. Shyong, D. Frankowski, J. Riedl, Do you trust your recommendations? an exploration of security and privacy issues in RS, in: International Conference on Emerging Trends in Information and Communication Security, Springer, Berlin, Heidelberg, June 2006, pp. 14−29.

[30] S. Badsha, X. Yi, I. Khalil, E. Bertino, Privacy preserving the user-based RS, in: 2017 IEEE 37th International Conference on Distributed Computing Systems (ICDCS), IEEE, June 2017, pp. 1074−1083.

[31] X. Liang, J. Tian, X. Ding, G. Wang, A risk and similarity aware application RS, J. Comput. Inf. Technol. 23 (4) (2015) 303−315.

[32] S. Badsha, I. Vakilinia, S. Sengupta, Privacy preserving cyber threat information sharing and learning for cyber defense, in: 2019 IEEE 9th Annual Computing and Communication Workshop and Conference (CCWC), IEEE, January 2019, pp. 0708−0714.

[33] J. Wang, A. Arriaga, Q. Tang, P.Y. Ryan, CryptoRec: Privacy-Preserving Recommendation as a Service, 2018 arXiv preprint arXiv:1802.02432.

[34] V.N. Gadepally, B.J. Hancock, K.B. Greenfield, J.P. Campbell, W.M. Campbell, A.I. Reuther, RS for the department of defense and intelligence community, Linc. Lab. J. 22 (1) (2016).

[35] J.A. Golbeck, Computing and Applying Trust in Web-Based Social Networks (Doctoral dissertation), 2005.

[36] M.M. Azadjalal, P. Moradi, A. Abdollahpouri, M. Jalili, A trust-aware recommendation method based on Pareto dominance and confidence concepts, Knowl. Base Syst. 116 (2017) 130−143.

[37] N. Polatidis, E. Pimenidis, M. Pavlidis, H. Mouratidis, RS meeting security: from product recommendation to cyber-attack prediction, in: International Conference on Engineering Applications of Neural Networks, Springer, Cham, August 2017, pp. 508−519.

[38] F. Ortega, J.-L. Sánchez, J. Bobadilla, A. Gutiérrez, Improving collaborative filtering-based RS results using Pareto dominance, Inf. Sci. 239 (2013) 50−61.

[39] R. Burke, B. Mobasher, R. Zabicki, R. Bhaumik, Identifying Attack Models for Secure Recommendation. Beyond Personalization, 2005, 2005.

Evolving cloud security technologies for social networks

Patruni Muralidhara Rao[1], Pedada Saraswathi[2]

[1]*School of Computer Science and Engineering, Vellore Institute of Technology, Vellore, Tamil Nadu, India;* [2]*Department of Computer Science and Engineering, GITAM Institute of Technology, Visakhapatnam, Andhra Pradesh, India*

1. Introduction

The rapid growth of social networks (SNs) and cloud services has led to the enhancement of online users in which the people can transform and access information. With the anonymous increase of social networking websites with additional applications, these websites have gained huge profits with the growth of online users. Consequently, a greater number of users are using SNs, which can increase various users to on online SNs. Since the storage of data can be increased exponentially, the storage of these data can be complex and difficult. As a solution, the cloud provides storage as a service to store this huge amount of data. Since additional personal information is stored on social networking sites, this violates privacy concerns. However, researchers have addressed several security issues in cloud-based SNs that violate confidentiality and privacy.

Of late, the evolutions of cloud computing (CC) models enable various types of users to store a large quantity of data in a flexible and scalable manner. Despite several advantages of using cloud as a service (CaaS) to store the data, various security-related issues have been identified that take down the performance of CaaS. Cloud technologies face several security challenges in both hardware and software over the internet, namely, privacy, integrity, availability, authorization, and confidentiality [1] of the data over the cloud. Subsequently, weak network traffic encryption across the network may also lead to the influence of confidential information. Data access can be done anywhere on the cloud since one can contact the information everywhere on the cloud; data access is also a major security-related issue in cloud-based networks. There are some cloud security vulnerabilities including denial of service (DoS) attack and distributed denial of service (DDoS) attack that assault on a computing environment. Data integrity must be maintained across the network to avoid data insertion, modification, and fabrication, so data might not be accessed by other users. Data authorization maintained across an environment helps to secure transmission of information [2].

Security in IoT Social Networks. https://doi.org/10.1016/B978-0-12-821599-9.00008-X
Copyright © 2021 Elsevier Inc. All rights reserved.

Over the past decades, we have seen diversion in computing mechanisms including communication, processing, and storage. As part of recent technologic advancements, various overwhelming dedicated connection caskets have been deployed everywhere. These advancements may lead to connecting with high-end mobile devices and wireless communication technologies made available with SNs. There is a need to protect the users from various security concerns including trust and autonomy, which demand control over cloud applications, data, and services away from central nodes. Therefore, edge computing (EC) can be an innovative pattern that has control over CC applications and services on the edge of network (EoN). An edge can be a wearable, mobile, or any user-controlled device in which EC can allow users to take up the control over personal information, using fewer resources and reduced response time to make fast social networking communications. In this chapter, we provide a framework that examines the security and privacy issues in the cloud-based SN.

1.1 Motivation

The stimulating element behind this chapter is to furnish a summary of advanced security and privacy issues due to the intrinsic usage of online social networks (OSN) using CC. To smoothen communication, every individual is necessitated to use advanced technologic facilities. For instance, social media can be one of these types of communication that has both positive and negative consequences to the users. SNs make information sharing more flexible and create faster communication. Subsequently, SNs establish the rapid developments of globalization as a reality and provide the ability for their users to express views. Also, they open doors to international relationships for both social and business relations and are the easiest way to communicate with each other using SNs at any time across the globe. Although, SNs have several disadvantages, among which the main issue is security and privacy. In this chapter, various issues related to SNs are discussed along with the recommendations used for the specified issues.

1.2 Background

An ever increasing array of cloud-based services in the new era has led to the development of cloud technology like CaaS, where the users can store a vast amount of data, and focus is gradually increasing to use the cloud as an environment that gives a robust and reliable performance. This increase in the number of users in the cloud results in the collection of a large amount of data and information from cloud services. In the traditional era, a cloud has been efficient, scalable, and effective, but in the emerging cloud service a single cloud with an extension of service stores data within a cloud. Most of the cloud services use online transaction processing (OLTP), which can operate and halt under high latency conditions. Moreover, a high volume of data used on a part of the network increased the load on websites such as Google, Facebook, etc. Rather than fetching the data, distributing a large amount of information from the client to the server decreases the response time to access the data. The rapid growth of SNs and the emerging popularity of online communication made every user's sensitive information available online. Most of the data shared on OSNs is insensitive, but some users may circulate their sensitive private information. However, publicly available sensitive data can be accessed easily, which leads to the leakage of privacy. Consequently, freely available data can be traced very fast, and the adversaries can be coupled to extract sensitive information.

In a cloud computing environment (CCE), privacy requirements are not properly defined, and proper encryption methods do not address privacy protection. Privacy requirements should be properly prioritized in CCE, so a person can easily access the information on a particular cloud, so privacy leaks can be properly distinguished from the information that is used in a cloud. Due to limited prospects provided by a cloud to the user's privacy, security plays a prominent role. Security is a critical viewpoint and continues to be problematic and must be considered according to privacy requirements. Although the hybrid cloud is a combination of both public clouds and private clouds, the private cloud ensures information security and privacy. This new era cloud computing also came a public cloud that ensures security by providing Security Assertion Markup language (SAML) for authentication and security. Cloud computing also differs from different computing techniques such as grid computing, utility computing, and edge computing, virtualization, client-server, so the development of cloud computing can be done in a stepwise manner. Related to security in the cloud, infrastructure, storage, and authorized access play a very prominent role in providing high security with more transparency.

Many organizations are moving from cloud computing to edge computing, to gain a good insight and work with small edge cloud, by focusing on information that resides on the cloud. Cloud computing is not that efficient when data is produced at the edge of the network and data processing is difficult and slow. As data is increasing more at the edge of the network, it is efficient to use edge computing to manage the data processing and speed of the network [3]. To define the programmability of edge computing, stream computing is more efficient, so a series of functions can be applied to the data along with the data propagation path. The computing can be defined in such a way that data can be processed and distributed across generated devices, with edge nodes residing within a CCE. In edge computing, most of the computing stream can be done at the edge rather than a centralized cloud, and it helps to know how the data propagation can be done at the edge of the cloud. Edge computing is a decentralized model that allows computing, storage, and service from cloud to edge devices such as smartphones, gateways, or routers. As mobile cloud computing enhances the capabilities of mobile devices, edge computing also extends the capabilities of storage, memory, performance, and processing power of data. Edge computing is assigned with small data centers in such a way that storage power will be greater. Service availability is higher in edge computing as connected devices in the edge environment do not need to wait for a highly centralized device. Moreover, EC can be a new computing paradigm that yields enhanced IoT data processing mechanisms, which enable data to be processed at or near the devices themselves. A local computer or server can be established as an edge node to process near the IoT device before sends it to the actual server or datacenter [4].

Finally, all the edge nodes will send the received data to the cloud storage. Mobile edge computing allows edge computing to be accessible at mobile services with low latency, and the improved version 5G makes mobile devices at low latency with higher bandwidth among services distributed across the network [7]. Social sensing edge computing (SSEC) became an important paradigm that can take measurements from the real world such as humans or machines. The working SSCE is based on the edge layer, service layer, and edge server layer. The edge layer is responsible for providing computational tasks, sensing, and storage near a device, and the edge server layer acts as an interface to all the users who are interested in using a particular application, The service layer acts as service between edge and server (intermediate layer), which comprises the gateway, bridges, cloudlets, and routers. The main importance of SSEC is that it is user-friendly and more flexible, with different system architectures. Social sensing entertains high privacy and security as the data is originated only on private edge devices [8]. More and more services are pressed from the cloud to the edge of the network. Moreover, handling data at the edge of the cloud increases processing speed, storage capacity, and better bandwidth utilization (Table 8.1).

1.3 Privacy protection and data integration

Generally speaking, privacy has several meanings depending upon the scenario and situation, and the intensive level of the privacy depends on the context of shared information. The first perspective of privacy is to provide the decisive value of the data to be saved to resist the contextual truthfulness of the SN's user records [9]. The data collected from the SN for the analysis process can be unintentional and mostly irrelevant. But it may be linked to the concealed activities of the user's information that may include religion or business or political associations (Fig. 8.1).

Table 8.1 Acronyms used.

Acronym	Abbreviation	Acronym	Abbreviation
OSN	Online social networks	EoN	Edge of network
CC	Cloud computing	OLTP	Online transaction processing
SN	Social networks	CCE	Cloud computing environment
CaaS	Cloud as a service	SAML	Security assertion markup language
DoS	Denial of service	SSEC	Social sensing edge computing
DDoS	Distributed denial of service	EC	Edge computing

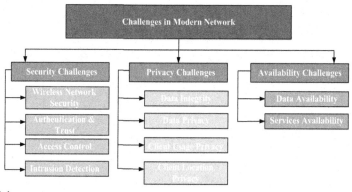

FIGURE 8.1

Various primitive challenges in advanced networks.

There is no doubt that security and privacy play a vital role in OSN. But more than 50% of SN applications do not spend adequate resources to enable security and privacy. To build security and privacy systems for these applications, the following security requirements are essential [10]. Although, the main aim of the EC paradigm is to accomplish various tasks, several security features and high performance are needs to be considered for the EC to ensure confidentiality, authentication, and integrity of different types of data. Generally speaking, as the OSNs are open to every mobile user, they are highly prone to physical attacks. Moreover, data transmission can be done via a wireless communication medium, by utilizing the wireless communication channel to interlink the mobile devices. Subsequently, this scenario can make the system highly vulnerable to various attacks including DoS, spoofing, sniffing, user impersonation, communication interception, and many more [11]. Because of the CC paradigm, the data storage and deployments of multiple user applications are major security risks. However, standard security mechanisms are available to secure the user's data. Although, these are becoming very complex in the EC environment since an increased number of related security issues in terms of traffic are carried over nodes. For example, an attacker can deploy malignant applications that can employ vulnerabilities causing damage or slowing the processing. Resultantly, it could lower the quality of service of the network [12] (Table 8.2).

2. Evolution of computing platforms in social networks

2.1 Evolution of cloud services

An ever increasing range of cloud-based services in the new era has led to the development of cloud technology like the CaaS, where the users can store a vast amount of data, and focus is gradually increasing to use the cloud as an environment that

Table 8.2 Security problems and potential solutions in OSN for mobile users.

Security problems	Requirements	Potential solutions	Protocols
Unauthorized access (authentication and authorization)	Secure key establishment	Cryptographic schemes and random key distribution	EAP, PAP, Kerberos, TACACS
DoS	Availability	Intrusion detection system and data redundancy	OSPF, BGP
Compromised node	Resistance to node compromise	Node tamper detection	LEAP
Message disclosure	Confidentiality and privacy	Link encryption and access control	SNMP
Message modification	Integrity and authentication	Secure hash key and digital signatures	SHA, MD5
Intrusion and vicious activities	Secure group management, intrusion detection system	Secure group communication intrusion detection system	PIDS, AB-IDS
Routing challenges	Secure routing	Secure routing protocols	SEAD, DSDV, SEF

gives a robust and reliable performance. This increase in the number of users in the cloud collects a large amount of data and information from cloud services. In the traditional era, a cloud was efficient, scalable, and effective, but in emerging cloud service a single cloud with an extension of service stores data within a cloud. Most of the cloud services use OLTP, which can operate and halt under high latency conditions. Moreover, a high volume of data used on a part of the network increased the load on websites such as Google, Facebook, etc. Rather than fetching the data, distributing a large amount of information from the client to the server decreases the response time to access the data (Fig. 8.2).

In a CCE, privacy requirements are not properly defined, and proper encryption methods do not address privacy protection. Overwhelming the privacy requirements should be properly prioritized in CCE, so a person can easily access the information on a particular cloud, so privacy leaks can be properly distinguished from the information that is used in a cloud. Due to limited prospects provided by a cloud to the user's privacy, security plays a prominent role (Fig. 8.3).

Security is a critical viewpoint and continues to be problematic and must be considered according to privacy requirements. Although the hybrid cloud can be a combination of both public and private clouds, a private cloud ensures information security and privacy. In this new era, cloud computing also came with a public cloud

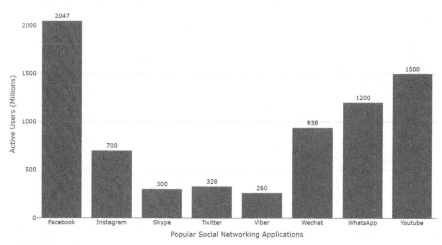

FIGURE 8.2

Popular SNs with their active users in millions.

that ensures security by providing SAML for authentication and security. Cloud computing also differs from different computing techniques such as grid computing, utility computing, and edge computing, virtualization, and client-server, so the development of cloud computing can be done in a stepwise manner. Related to security in the cloud, infrastructure, storage, and authorized access play a very prominent role in providing high security with more transparency (Fig. 8.4).

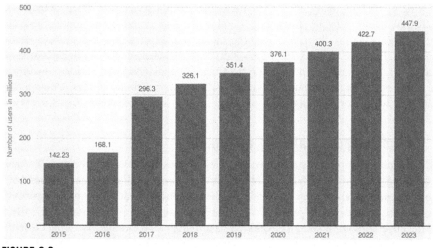

FIGURE 8.3

Number of social network users in India in millions.

Statista; Statista Digital Market Outlook.

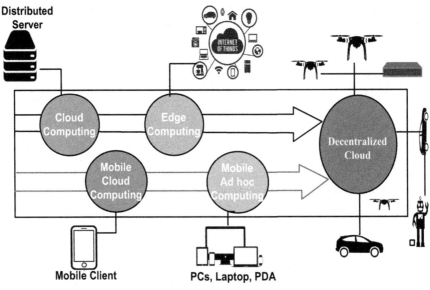

FIGURE 8.4

Evolution of computing paradigms.

2.2 Cloud computing paradigm for SNs

Over the last decades, various computing paradigms have been serving for the betterment of SN platforms. Similarly, in the recent past, cloud computing emerged as a leading efficient platform for storing and processing massive amounts of data. It is widely used in many industries including healthcare, real estate, education, industrial manufacturing, financial, social networking, and many more. Several organizations deploy their data into the cloud instead of deploying to their local machines. Because cloud computing has diverse characteristics, including elasticity, scalability, pay-per-use pricing, infrastructure management, various security services, etc., cloud computing works based on a distributed model. Though, it primarily works for the outsourcing of data storage and processing (Fig. 8.5, [5]).

2.3 Edge computing and fog computing

EC is a new computing paradigm that yields the enhancement of the IoT data processing mechanism, which enables data to be processed at or near the devices themselves. A local computer or server can be established as an edge node to process close to the IoT device before sends it to the actual server or datacenter [4]. Finally, all the edge nodes will send the received data to the cloud storage. With the current statistics, 92% of data workload in the following years is expected to have a control

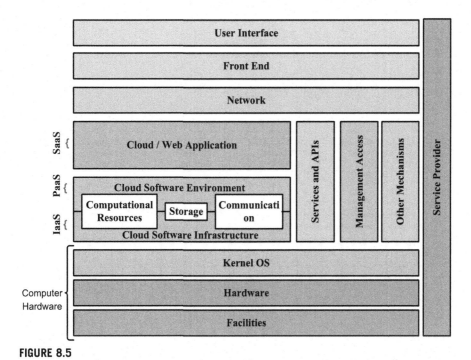

FIGURE 8.5

Cloud computing paradigm for online SNs.

layer of fog computing (FC) as highly demanded. This can serve devices near the end to facilitate mobility and data locality. Basically, in 2012, Cisco introduced the concept of FC, and it is getting more attention for the development of smart applications from academia and industry [7]. Subsequently, it has opened the door to various researchers to enhance performance and security. The prime goal of deploying FC at the edge is to get the benefit from various services including storage, bandwidth, computation, and processing the real-time data for the adjacent nodes in one-hop fashion to minimize latency [8,9]. Moreover, FC can be another outer layer of distributed network computing that is closely associated with a cloud and the IoT. Generally speaking, fog node provides the missing link for what the data needs to be sent to the cloud storage (Fig. 8.6).

Fog and edge could conversely work for cloud and IoT because they are involved at an intermediate level of processing and storage infrastructure. However, the major difference between fog and edge is the location of the computing infrastructure. In fog, local area networks work as a gateway, whereas the edge can do the computing on the device itself using smart devices like programmable automation controllers [6].

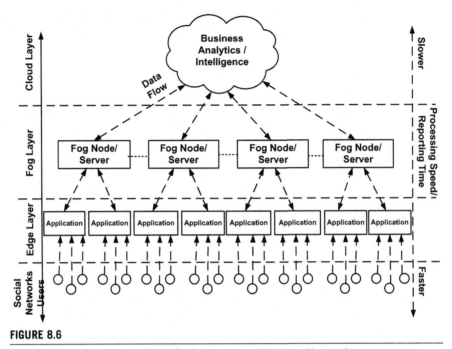

FIGURE 8.6

Interaction between the cloud and fog/edge with social networking nodes.

3. Access control and security policies in social networks

3.1 Architecture and use case model

Of late, as there is a rapid growth of active users on OSNs, protecting their personal information with the right security and privacy mechanisms is highly demanded. To secure data on SNs, various security mechanisms have been proposed to maintain security for specific SNs by remodeling and reorganizing the visualized graphs, among which "anonymization" can be the premier algorithm-based mechanism widely used for privacy protection. Also, various decentralization and probabilistic-based algorithms have been proposed to depreciate the weaknesses of anonymization. Besides, these algorithms have concrete limitations concerning computational complexity and robustness. Nonetheless, SN platforms can be integrated with third-party applications. These app developers provide interfaces to access people's personal information (Fig. 8.7).

Various public-related organizations including banking, transportation, insurance, and telecommunications combine data proficiently by crawling billions of user profiles, which accordingly puts the user's information at risk. To protect confidential sensitive information, data anonymization on the network is needed [13]. Thus, to ease the mentioned issue, several third-party algorithms are prescribed to

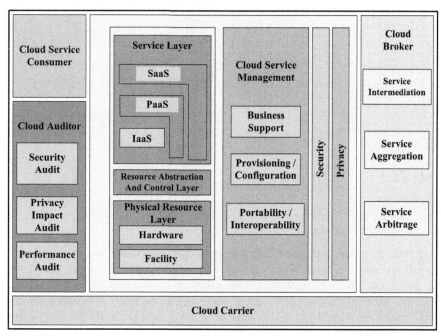

FIGURE 8.7

Architecture for cloud security in SNs.

stand with privacy policies, and many more third-party applications are being supported by OSNs to be performed by the users, in which they can limit the transfer of a user's personal information to external application vendors.

The extensive growth of OSNs makes their structure dynamic, in which the evolution of the network brings privacy violation of the user's data. To generalize the network modeling, link prediction algorithms are being utilized through graphs. To provide privacy without information loss, a group-based anonymization algorithm can be produced. Therefore, this section focuses on security and privacy attacks and various solutions being implemented to minimize the risk from these attacks as researched. Furthermore, Fig. 8.8 depicts the categorization of various properties of OSNs including large-scale implementations, network-based clustering, and degree dispensation [14]. As depicted in Fig. 8.9, these are again classified into four subcategories. They are static, dynamic, diffusion influence models, and user behavior-driven models.

3.2 Issues

Initially, privacy can be defined as any right to control the user's personal information and how it is used. In connection with OSNs, the four categories of privacy concerns are depicted in Fig. 8.9.

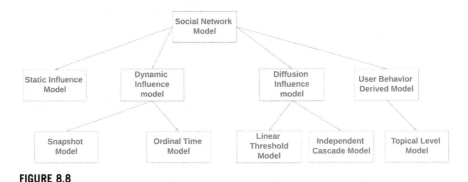

FIGURE 8.8

OSN network model.

A. User-based limitations:

In this, the primary users of OSNs have various limitations and flaws. Due to these flaws, they share their personal sensitive information on public platforms like Facebook, Twitter, Instagram, and many more opening doors to privacy breaches. The OSN users interestingly post massive amounts of information on social media on several occasions without being concerned with the short-term and long-term results.

B. Design flow:

This is one of the weak privacy controls of design limitations and flaws such as SN attacks and cloning attacks. Resultantly, the inability of SNs may lead to mistrust of propagation of the user's data. Due to the OSN's inflexible privacy policies, various attackers can create fake accounts easily and then impersonate someone's data. Consequently, adversaries can use the user's personal information including pictures and videos on any fake profile to gain the trust of their friend's connections. Therefore, these fake accounts can enter into their limits of trust factors. For instance, Facebook has given the advantage of recovering suspended accounts, even though those benefits can be revoked if the adversaries associated with the victim's account

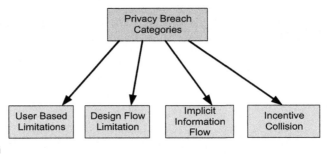

FIGURE 8.9

Privacy beach categories.

are found with the newly created fake account. Resultantly, disconnecting the victim's login credentials is associated with the present account. These kinds of attacks are known as permanent takeover attacks [11]. Thus, Facebook banned the distributed network of already used login details including username and email address with newly created accounts. Adding targeted recipients and further deactivating their accounts comes under the scenario of deactivated friend attacks [11].

C. Implicit information:

Implicit information leaks are those in which one leak leads to another information leak. According to the Facebook study of over 1.47 million accounts, Ref [15] determined that 1.5% of users only reveal their age. For instance, in high school friend connections of Facebook users, one can estimate the ages of approximately 85% of these users with mean error ± 4. Subsequently, the user's activity related to their videos, pictures, page likes, and clicked ads leads to implicit information leaks. Resultantly, adversaries can misuse this implicit information leak.

D. Incentive collision:

Almost every SN can be supported by the revenue created by services from advertisements. This leads to creating conflict between the user's access rights and the advertiser's right concerning the service provider. Most of the service providers depend on advertiser income for the proper functioning of the services and profit-making has been demanded for every hour. Thus, various issues are unresolved between the user's and the advertiser's perspectives. Resultantly, these impacts lead to privacy breach from the service provider side.

4. Cloud security technologies

The research on SNs is still finding ample qualitative and quantitative analysis of trustworthiness and security. In connection with the newly developed social networking applications and novel features, various researchers have proposed novel directions with measurable evaluations and security of SNs by characterizing transparency, authenticity, availability, and quality assurance. Fig. 8.10 depicts the top-down approach from evidence-driven signals for SN platform security and trustworthiness to build a trusted, secure SN ecosystem.

4.1 Cloud computing paradigm

CC is one of the foundations for the next generation of computing. It comprises computing services including data, storage, software, processing power, and applications that are distributed through the internet. In this, the services are completely served by the provider, and the client just needs a lightweight personal device with internet access. In this paradigm, cloud deployments can be done either onsite by the organization or offsite including SkyDrive, Google Drive, S-Cloud, iCloud, and Amazon Cloud Drive [41].

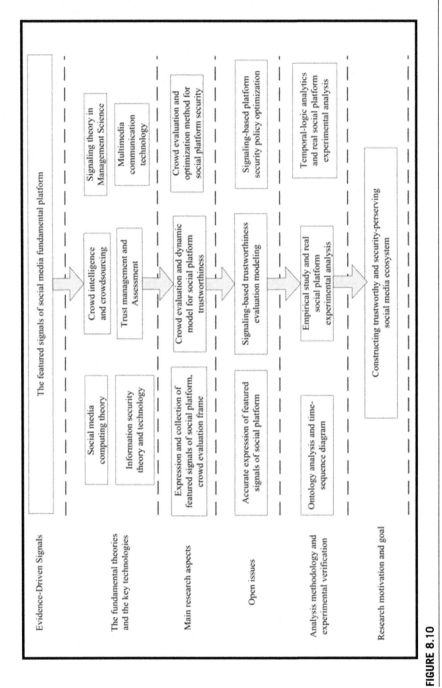

FIGURE 8.10

Directions of trustworthiness and security in social networks.

4.2 **Edge computing paradigm**

Fig. 8.11 depicts the communication between the users and the edge computing network. Here the user requests services from local or remote servers. Also, the end users can communicate with each other to calculate the conversation key for specific usage. As communication is established via public channels [6], it is in danger of various security attacks.

4.3 **The threat model for SNs**

In the recent past, Laorden et al. [19] introduced a threat model for SNs to study threats and vulnerabilities. In this, they proposed a Circle-of-Risk (CoR) model for SNs threats and vulnerabilities. This CoR comprises various modules including assets, attacks, threats, vulnerabilities, and several countermeasures. This CoR describes various security features of the threat model. But, this model cannot satisfy various questions including where the threats and vulnerabilities come from, what are the countermeasures, and what are the system components. Therefore, this section describes various security threats, attacks, and recommendations (Fig. 8.12)

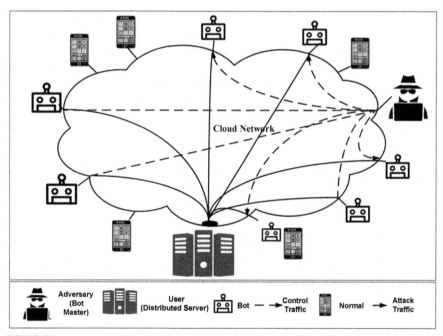

FIGURE 8.11

DDoS attack scenario for cloud computing using mobile networks.

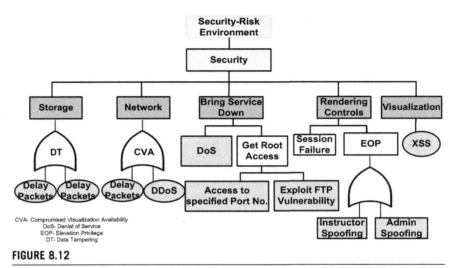

FIGURE 8.12

Privacy and safety security attack tree.

4.4 Privacy and security threats in social networking applications

Today, massive changes are being done in human's personal and business lives with the help of SN tools. However, SNs play a vital role in connecting human's social and business lives. Today, millions of users regularly use SNs in which they open doors to attackers. Resultantly, it leads to prominent risks about security and privacy threats. These can be classified into two categories [14]. They are classical and advanced threats. First, classical issues are online threats that make SNs vulnerable. Second, advanced threats are exclusively related to online users in which some of the known threats can be described as follows [2].

Malware: It is a malicious software that is developed to enter into another's computer to access their contents. This attack on SNs can be easier than other online services due to the construction of OSNs and other contacts among the users [21].

Spam attacks: These attacks are rapidly growing and becoming a challenging issue to SNs, in which spam greetings are unwanted messages. However, spam will come through a post on a wall or spam instant messaging in SNs. These attacks are more dangerous in SNs than traditional email spam attacks. Spam consists of notices or malicious links that may cause phishing. Particularly, spam attacks come from fake profiles that are normally spread from a profile created in the name of a well-known person [23]. Moreover, most of the spam is distributed through compromised accounts. Thus, spam-filtering techniques are required to detect malicious messages or URLs in a message. Similarly, one should filter them before delivery to the targeted web sites [38].

Phishing attacks: These attacks are one of the most common attacks under social engineering. Phishing attacks have become of the major issues in web security. In this, the vulnerable user receives a phishing link targeting them that can be redirected to an illegitimate, vicious website. Resultantly, the user's account will be compromised upon entering their credentials into the malicious website [22].

XSS (cross-site scripting attacks): These are some of the most common computer security threats. They are found in web applications, in which they can enable malicious adversaries to inject client-side script code into the specified web pages that can be seen by other users [20].

Sybil attacks: This is a fundamental attack in which the attacker creates different accounts and joins the targeted network to gather personal information and group information including voting, likes, shares, etc. [25].

De-anonymization attack: In these attacks, an attacker reveals the integrity of a SN user from an advertised anonymized graph. Recently, a few researchers have modeled a $re - identification$ mechanism to identify users from published anonymized graphs [24].

Clickjacking: It is widely known as a user interface redress attack. In these attacks, adversaries particularly use a malicious technique to induce users to click on something that is not what they intended. In these attacks, posting spam posts into the SN user's timeline is considered, in which they manipulate SN users, and in turn, they ask for likes to links without their notice [26]. With these attacks, an adversary can also use a SN user's hardware, for instance, camera, voice recorders for voice recording, GPS for location information, etc.

Fake profiles: Today, it is one of the most common attacks in the SN. In this attack, an adversary creates a fake SN profile with fake details and credentials then sends messages to valid users. Similarly, upon receiving the response from them, it sends spam messages to the user. The main aim of these fake profiles is to gather massive information from SN user's private data, which is accessible only to their own friends and aimed to spread spam messages [27].

Identity theft attacks: It is a kind of profile-cloning attack. In this, an attacker performs profile cloning using theft credentials from legitimate users, creating a new fake profile utilizing these credentials. And these credentials can also be used within various other networks. These types of attacks are also called identity clone attacks (ICA). In this, an adversary uses the beliefs of the cloned user profile to gather valuable content from their known peers to encompass online fraudulent activities [28].

Inference attacks: These attacks target a large amount of user's data including name, age, relationship, profession, activities, etc.

Cyberstalking: This is a typical SN attack, in which adversaries provoke the individual or group users through various SN platforms. Cyberstalking can be primarily used for monitoring, ICA, threats, plagues, and solicitation for sex, etc. In Ref. [29], an online survey conducted on women's experiences with cyber harassment was described. In this, about 300 participants of various SNs were taken for the survey. Similarly, nearly 59% were students at college, and nearly 10% of women received pornographic messages from unknown users, nearly 20% of women continually received sexual harassment messages on the internet, and almost 33% of women experienced cyber harassments [30].

4.5 Recommendations

As SNs have various security and privacy issues, these can be controlled by using precautionary methods. An adversary can use these security and privacy concerns of social media because of the negligence of the user's activities. Consequently, the contents shared by these SN users via their friends can be authorized to go in the wrong way in a different context. Also, by using identification techniques of a profile, the shared contents involved with other public datasets may be reconstructed. Resultantly, it may lead to the disclosure of personal privacy. Various safeguarding mechanisms have been introduced against privacy issues that can be provided through the privacy settings that can be controlled by SNs. However, the impact of these privacy settings is not efficient due to the design perspectives of the agreements between the user and OSN. This user agreement allows for collecting more information from the users rather than being concerned with protecting their privacy [11]. Thus, various recommendations for safeguarding user content and privacy from unauthorized access have been studied, and they are described as follows.

Personal information: Today, most SN users use third-party applications to share content. Once the SN user's private contents are distributed through these applications, there is no further assurance that the content can be safe. Thus, it is suggested not to share private data. Although the SN user might know the value of privacy, these SN privacy policies can frequently change to create confusion about the shared contents of the SN user [16].

Location information: As most of the mobile applications gain user's location information, this can be used to send to third-party organizations majorly for a commercial purpose that leads to privacy issues. Every SN user is recommended not to use apps requiring location information, but often they may share location information with their posts. Adversaries may take advantage of misusing this location information by knowing your present place. Thus, SN users are not recommended to disclose their location information.

User privacy settings: According to recent statistics, more than 80% of OSN users do not know the privacy of their profile nor did they check their privacy settings. However, SNs widely offer various levels of access control for the data through custom settings to hide the valuable contents from unauthorized access points. Most of the OSN users keep the default privacy and security settings [15]. In this connection, it is recommended to use the custom privacy settings and use the largest advantages of privacy protection standards provided by their SN platforms. Also, it is further recommended to revise their privacy settings thoroughly as the SNs release updates that automatically change their privacy settings.

Third-party applications: As most of the third-party applications are growing, massive security and privacy concerns are raised because their code is deployed outside the SN's control. This can primarily stop SN users from controlling the application activities and predicts vulnerable penetration. Subsequently, the data is being transmitted out of the SNs, where the usage of the user's content cannot be in the control of the SN users. To safeguard personal information from SNs, it is recommended to *uninstall* third-party applications.

Antivirus and spyware: SNs are widely emerging platforms to enable communication between individuals where content sharing can be easily done [17]. The reliable use of SN platforms allows massive content distribution; subsequently, malware dissemination has also grown exponentially [18]. Malware can be a vulnerable software used for interrupting user activities, to illegally collect information, and to gain unauthorized access for private data. Thus, SN users are recommended to install antivirus and antispyware to encounter these issues.

5. Manifesting mechanism

5.1 Security modeling

In the recent past, cybercrimes have been rapidly increasing through the extensive use of SNs, resulting in harmful consequences, even the loss of life. In this connection, several types of threats have been identified. These SN threats are categorized into three groups to provide the process of finding a suitable security model. According to Ref. [31], model SN threats represent primarily three groups depicted in Table 8.3. Similarly, the classification of these threats can be described in Table 8.4 based on these three groups, in which some threats may include other groups also.

Generally speaking, many cyber-attacks mainly target a particular group of people in SNs based on gender, religion, political affiliation, profession, ethnicity, etc. They can send posts according to the group based on the target. Those posts can affect people's reputation, the business of the organization, and harm an individual through real-life harassment (Fig. 8.13).

Several terrorist groups use SN platforms to share their views, ideas, publish extreme opinions, recruit people who can fight against the system, and communicate to them through SNs. In the recent past, cyberterrorists have become very dangerous to society, in which they widely have used online SN platforms. According to Statista Online Platform [31], the *ISIS terrorist group* flourishes on Twitter OSN, where more than 12,000 accounts were created. Also, these SNs can be used to share specific malicious content including images, videos, and flash files. For instance, information leakage is another kind of privacy leakage that allows SN users to share their information with friends, friends to friends, friends to groups, groups to groups, and other users. According to recent a Facebook report issued by Cambridge Analytica, more than 50% of user's personal information was accessed without their notice [32,40].

Table 8.3 Social networking threats based on groups.

Group	Threats	Description
I	Account based	Related to user's authenticity and access control
II	URL based	They lead to external resources, specifically the posts, in which they include URLs
III	Content based	Caused by shared content

Table 8.4 SNs threats classification based on groups.

Threat	Account	URL	Contents
Phishing and social engineering	✓	✓	
Fraud and spam posts	✓	✓	✓
Malware propagation		✓	
Cross-site scripting		✓	
Clickjacking	✓	✓	
Identity theft	✓		
Data leakage	✓		✓
Cyberbullying	✓		✓
Malicious links		✓	
Hate crimes			✓
Cyberterrorist	✓	✓	✓
Revenge porn	✓		✓
Inappropriate contents		✓	✓

To facilitate security modeling for the aforementioned SN threats based on grouping, many researchers have introduced various security modeling techniques. Ref. [33] devised a classification model to detect fake accounts in Twitter. They worked on around 22 parameters considered to identify based on the existing model. Later, with help of feature selection technique, they reduced from 22 to seven efficient parameters. Similarly they applied classification technique for five algorithms, as various other scenarios to calculate accuracy of the proposed model. Also, they assured that their model can produce 99.9% accuracy, but time delay was not mentioned. Subsequently, authors in Ref. [34] employed map-reducing mechanism using pattern recognition technique to detect fake Twitter accounts. In this, they used

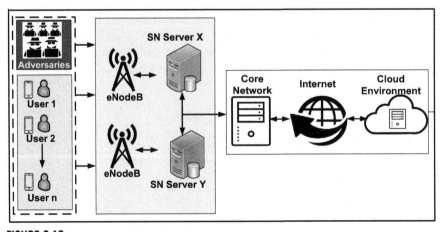

FIGURE 8.13

A generic attack scenario in social networking.

33 parameters, and the analysis depicts the fake accounts were created based on batches with particular intervals, targeted to post spam tweets. This mechanism can detect fake accounts without elaborate analysis of Twitter content. However, this mechanism cannot be suitable for real-time scenarios, where this mechanism needs to monitor an account's behavior. Also, they ensure their contribution includes facilitating a standard data set that can be used further to research identifying effective spam detection.

Consequently, Ref [35] created a fake account detection mechanism utilizing *feature − based detection*a feature-based detection mechanism with Machine Learning model classifiers. They insured that their results have predicted nearly 95% of appropriately categorized accounts. The authors in Ref. [36] devised an open-source protection mechanism to detect fake profiles in SNs. Their recommended mechanism uses user ranking operations based on user level that activates and represents values in the graph as weights. In this mechanism, they require nearly about 70 h to detect 20,000 fake accounts from the lowest data sets, which is a comparatively time-consuming process and not suitable for real-time applications.

Thereafter, various cryptographic schemes have been proposed to withstand these threats. Ref [37] facilitated an accessible security application driven by public-key *cryptography* and allowed the end-end-encryption over distinct devices. Ref [38] introduced an automated key exchange scheme mobile-based application. It includes security channels to facilitate the authenticity and confidentiality of the users and data, respectively. However, these applications cannot support some standard mechanisms to publish public keys, digital signatures, and user's expertise to handle keys and cryptographic algorithms. Thus, overheads and latency were not confirmed to be sufficient for real-time applications.

Al-Turjman et al. [39] modeled a hybrid trust evaluation framework used for trust measurement in an e-commerce SN. Sometimes, it is a challenging task to identify trusted SN users instead of users with high profile rating and more reputation. This model facilitates a secure and reliable platform for e-commerce management to facilitate trust between the users in the SN.

5.2 Access control enforcement for fog/edge computing

Personal information can be largely collected and deeply processed in the FC. But, it cannot be adopted efficiently without proper privacy and security mechanisms. The FC may face several classical issues inherited from the classical CC paradigm. Alongside this, it has been proven that it will be inflicted with unique vulnerabilities such as man-in-the-middle, DoS, and DDoS attacks due to the adoption of fog nodes in the CC [12]. Thus, access control is very important to assure security, since it has a mechanism to allow or confine user's access to the authorized system. As a result, access control ensures authorized access for valid users and restricts attacks of unauthorized users. Also, it ensures resolving the security issues caused by fault operations of authorized users as well. However, in the traditional access control mechanism, the user's data can be stored in a trusted physical server. Subsequently, the trusted server checks for the adjured user's access privileges [13]. In the fog-enabled CC, the users and servers are deployed in distinct domains in which servers are untrusted because of diverse locations. However, in fog-enabled cloud computing models, using fog nodes, Cloud Service Provider's do not trust each other because they followed pay-as-you-go services on a pay-per-use basis. Of late, few researchers have addresses various access control issues, which they categorized in three ways.

I. If any user wants to access the storage and processing services, they need to authorize by fog- or edge-enabled cloud. And concerning policies must be used for control abstraction of the users' data and services.
II. Fog and cloud should have mutual access control.
III. To avoid side-channel attacks, we need virtual machine access control as well.

Access control in the FC is always an essential mechanism to maintain users' privacy and enhanced security. However, the resource constraints and design requirements of fog-enabled cloud environments are major challenges. While comparing FC with CC, we need to look forward to the following access control requirements to build a secure and effective access control mechanisms [39] (Table 8.5).

Table 8.5 Access control requirements in fog computing.

Requirements	Description
Latency	FC should concede access decisions in less time while offering limited computation cost, network foraging time, speed of policy decision, and offloading task time.
Efficiency	In FC, some nodes are enabled with ample resources and some of them are resource constrained. The efficiency of access control is a challenging aspect of any access control mechanism that could implement any policy. Subsequently, it may cause a delayed decision process, resulting in unacceptable latency to the other parts of the network.
Generality	Since systems and services are based on various tools enabled with software and hardware, generic API's are in high demand to deal with the existing API and protocols. Due to the heterogeneity of fog devices, there is a need for providing a common abstraction to the top-layer services and applications.
Aggregation	To reduce the latency, it should aggregate fog devices close to users because the user's data can be collected from various devices of distinct locations. When the data generated from the user is useless, however, intelligent aggregation should be done.
Policy management	It is an essential part of the access control model, which should have an ability to support invoking, releasing, creating, or deleting a policy.
Resource restriction	It is very much important because computation resources are limited at both the client and fog side.
Privacy protection	Since the FC has decentralized architecture, protecting the privacy of data is an essential requirement of fog access control.

6. Conclusion

In social networking platforms, security is a critical viewpoint and continues to be problematic and must be considered according to privacy requirements. However, the hybrid cloud can be a combination of both public and private clouds, where the private cloud ensures information security and privacy. In this new era, cloud computing also came with a public cloud that ensures security by providing SAML for authentication and security. Cloud computing also differs from different computing techniques such as grid computing, utility computing, and edge computing, virtualization, and client-server, so the development of cloud computing can be done in a stepwise manner. Related to security in the cloud, infrastructure, storage, and authorized access play a very prominent role in providing high security with more transparency. In this chapter, we have provided the state-of-the-art methods and security modeling for secure SNs that have been studied.

References

[1] D. Deebak, F. Al-Turjman, L. Mostarda, Seamless secure anonymous authentication for cloud-based mobile edge computing, Comput. Electr. Eng. J 87 (2020) 106782.

[2] S. Ali, N. Islam, A. Rauf, I.U. Din, M. Guizani, J.J. Rodrigues, Privacy and security issues in online social networks, Future Internet 10 (12) (2018) 114.

[3] J. Su, Y. Cao, Y. Chen, Privacy preservation based on key attribute and structure generalization of social network for medical data publication, in: International Conference on Intelligent Computing, Springer, Cham, August 2019, pp. 388–399.

[4] F. Xhafa, B. Kilic, P. Krause, Evaluation of IoT stream processing at edge computing layer for semantic data enrichment, Future Generat. Comput. Syst. 105 (2020) 730–736.

[5] A. Sunyaev, Cloud computing, in: Internet Computing, Springer, Cham, 2020, pp. 195–236.

[6] B.D. Deebak, F. Al-Turjman, M. Aloqaily, O. Alfandi, IoT-BSFCAN: A smart context-aware system in IoT-Cloud using mobile-fogging, Future Gener. Comput. Syst 109 (2020) 368–381, https://doi.org/10.1016/j.future.2020.03.050.

[7] G. Alonso Nuñez, Data processing handover in the multi-access edge computing setting, 2019.

[8] C. Cicconetti, M. Conti, A. Passarella, D. Sabella, Toward distributed computing environments with serverless solutions in edge systems, IEEE Commun. Mag. 58 (3) (2020) 40–46.

[9] E.W. Ayaburi, D.N. Treku, Effect of penitence on social media trust and privacy concerns: the case of Facebook, Int. J. Inf. Manag. 50 (2020) 171–181.

[10] W. Sun, Z. Cai, Y. Li, F. Liu, S. Fang, G. Wang, Security and privacy in the medical internet of things: a review, Secur. Commun. Network. 12 (2018).

[11] F. Al-Turjman, Intelligence and security in Big 5G-oriented IoNT: An Overview, Future Gener. Comput. Syst. 102 (1) (2020) 357–368.

[12] P. Zhang, J.K. Liu, F.R. Yu, M. Sookhak, M.H. Au, X. Luo, A survey on access control in fog computing, IEEE Commun. Mag. 56 (2) (2018) 144–149.

[13] C. Stergiou, K.E. Psannis, B.G. Kim, B. Gupta, Secure integration of IoT and cloud computing, Future Gener. Comput. Syst 78 (2018) 964−975.

[14] H. Kizgin, B.L. Dey, Y.K. Dwivedi, L. Hughes, A. Jamal, P. Jones, N.P. Rana, The impact of social media on consumer acculturation: current challenges, opportunities, and an agenda for research and practice, Int. J. Inf. Manag. 51 (2020) 102026.

[15] X. Liang, Security and privacy preservation in mobile social networks, 2013.

[16] A. Pradeep, S. Kasun, Securecsocial: secure cloud-based social network, vol. 1, World Scientific, 2019.

[17] B.B. Dedeoğlu, B. Taheri, F. Okumus, M. Gannon, Understanding the importance that consumers attach to social media sharing (ISMS): scale development and validation, Tourism Manag. 76 (2020) 103954.

[18] D. Koggalahewa, Y. Xu, An investigation on multi view-based user behavior towards spam detection in social networks, in: Pacific-Asia Conference on Knowledge Discovery and Data Mining, Springer, Cham, April 2019, pp. 15−27.

[19] C. Laorden, B. Sanz, G. Alvarez, P.G. Bringas, A threat model approach to threats and vulnerabilities in online social networks, in: Computational Intelligence in Security for Information Systems 2010, Springer, Berlin, Heidelberg, 2010, pp. 135−142.

[20] S. Lekies, K. Kotowicz, S. Groß, E.A. Vela Nava, M. Johns, Code-reuse attacks for the web: breaking cross-site scripting mitigations via script gadgets, in: Proceedings of the 2017 ACM SIGSAC Conference on Computer and Communications Security, October 2017, pp. 1709−1723.

[21] B.B. Gupta, S.R. Sahoo, V. Bhatia, A. Arafat, A. Setia, Auto fill security solution using biometric authentication for fake profile detection in OSNs, in: Handbook of Research on Intrusion Detection Systems, IGI Global, 2020, pp. 237−262.

[22] M.M. Ali, M.S. Qaseem, M.A.U. Rahman, A survey on deceptive phishing attacks in social networking environments, in: Proceedings of the Third International Conference on Computational Intelligence and Informatics, Springer, Singapore, 2020, pp. 443−452.

[23] M. Aldwairi, L.A. Tawalbeh, Security techniques for intelligent spam sensing and anomaly detection in online social platforms, Int. J. Electr. Comput. Eng. (2020) 10, 2088−8708.

[24] L. Fu, J. Zhang, S. Wang, X. Wu, X. Wang, G. Chen, De-anonymizing social networks with overlapping community structure, IEEE/ACM Trans. Netw. 28 (1) (2020) 360−375.

[25] T. Gao, J. Yang, W. Peng, L. Jiang, Y. Sun, F. Li, A content-based method for sybil detection in online social networks via deep learning, IEEE Access 8 (2020) 38753−38766.

[26] M. Niemietz, Analysis of UI redressing attacks and countermeasures doctoral dissertation, Ph.D. thesis, Ruhr University Bochum, April 2019, https://doi.org/10.13154/294-6394 (cit. on p.).

[27] M. Chatterjee, Detection of fake and cloned profiles in online social networks, 2019 (Accessed 22 April 2020).

[28] U. Can, B. Alatas, A new direction in social network analysis: online social network analysis problems and applications, Phys. Stat. Mech. Appl. (2019) 122372.

[29] S. Burke Winkelman, J. Oomen-Early, A.D. Walker, L. Chu, A. Yick-Flanagan, Exploring cyber harassment among women who use social media, Univers. J. Public Health 3 (5) (2015) 194.

[30] S. Jabbar, S. Khalid, M. Latif, F. Al-Turjman, L. Mostarda, Cyber security threats detection in internet of things using deep learning approach, IEEE Access 7 (1) (2019) 124379−124389.

[31] N. McCarthy, Infographic: ISIS is expanding its reach on Twitter, Stat. Infogr. (2018). Accessed Online: March 2020.

[32] Z. Tufekci, Facebook's surveillance machine, vol. 18, 2018. New York Times(Accessed 22 April 2020).

[33] A. ElAzab, Fake accounts detection in twitter based on the minimum weighted feature, World, 2016 (Accessed 22 April 2020).

[34] K.N. Güngör, O.A. Erdem, İ.A. Doğru, Tweet and account-based spam detection on Twitter, in: The International Conference on Artificial Intelligence and Applied Mathematics in Engineering, Springer, Cham, April 2019, pp. 898−905.

[35] M. BalaAnand, N. Karthikeyan, S. Karthik, R. Varatharajan, G. Manogaran, C.B. Sivaparthipan, An enhanced graph-based semi-supervised learning algorithm to detect fake users on Twitter, J. Supercomput. 75 (9) (2019) 6085−6105.

[36] X. Jiang, Q. Li, Z. Ma, M. Dong, J. Wu, D. Guo, QuickSquad: a new single-machine graph computing framework for detecting fake accounts in large-scale social networks, Peer-to-Peer Netw. & Appl. 12 (5) (2019) 1385−1402.

[37] Keybase, Keybase.io [Online], 2018. Available: https://keybase.io/ (Accessed: 21 April 2020).

[38] F. Al-Turjman, B.D. Deebak, L. Mostarda, Energy aware resource allocation in multi-hop multimedia routing via the smart edge device, IEEE Access 7 (2019) 151203−151214.

[39] F. Al-Turjman, Y.K. Ever, E. Ever, H.X. Nguyen, D.B. David, Seamless key agreement framework for mobile-sink in IoT based cloud-centric secured public safety sensor networks, IEEE Access 5 (2017) 24617−24631.

[40] S.K. Dasari, K.R. Chintada, M. Patruni, Flue-cured tobacco leaves classification: a generalized approach using deep convolutional neural networks, in: Cognitive Science and Artificial Intelligence. SpringerBriefs in Applied Sciences and Technology, Springer, Singapore, 2018.

[41] X. Yao, G. Li, J. Xia, J. Ben, Q. Cao, L. Zhao, D. Zhu, Enabling the big earth observation data via cloud computing and DGGS: opportunities and challenges, Remote Sens. 12 (1) (2020) 62.

Fake news in social media recognition using Modified Long Short-Term Memory network

Sivaranjani Reddi, G.V. Eswar

Department of Computer Science and Engineering, Anil Neerukonda Institute of Technology and Science, Visakhapatnam, Andhra Pradesh, India

1. Introduction

Nowadays, people are spending a huge amount of time viewing emails or browsing the internet or in sharing their status and spreading news and updates with their mobiles. Internet and social media are the means where people progressively tend to obtain political or entertainment news along with traditional news organizations. However, people are more inclined to use social media than traditional internet. There are many influential factors that motivate the users to attract them toward social networking services: (i) it is regular and up-to-the-minute, less expensive to dissipate as well as receive news on social media (SM) than the traditional news media (TNM), like periodicals including newspapers (NP), weekly and biweekly journals, or TV; and (ii) it is simple and straight to give remarks, criticize, express opinions, and discuss the news with friends or SM users. For example, In the United States, 62% of adults were addicted to reading news on a social net in 2016, whereas, in 2012, 49% were watching news on social sites. These statistics witness that social media is dominating the other major news-spreading media like newspapers and television. Despite the benefits delivered by SM, the value of news on an internet forum is poorer relative to old-style newscast organizations. However, due to the fact of lower cost, faster and easier news spreading over the internet community raises the scope to share volumes of false news, created and shared online, such as defaming a company's financial scenario and for party political gain. It was observed that more than 1,000,000 instances of fake news were tweeted before the presidential election. This may damage the name and fame of the concerned person. The extensive spreading of fake news creates a serious impact not only on the concerned individual but on society also. The impacts may include, first, that fake news can spoil the legitimacy of the news environment. The most popular fake news evidence was spread on Facebook during the president elections in the United States in 2016 compared to the most popular authorized news sharing media. Secondly, this spread fake news intentionally influences the users to accept false beliefs. In fact, some report shared in the media shows that forged accounts and social bots disseminated incorrect writings that were created in Russia. Third, it may change people's interpretation who respond naturally to the real news. To support this, some created ingenuine news that made people lose trust, created a lot of confusion, and inhibited their abilities to make a decision on which news is genuine.

Distinguishing fake news from genuine news poses numerous inspiring research problems. Though fake news itself is an old problem to nations, who have been using SM to publish information for eras, the increase of web-generated news on SM shows fake news is an influential force that is challenging the old-style news sharing mechanism. There are quite a lot of characteristics in this problem that make it exclusively challenging for computerized recognition. The challenges in automated fake news detection include, first, categorization of news into fake or genuine, where fake news intentionally aims to misguide readers. The news description of fake news is in terms of media platforms, and fake news tries to falsify truth through different etymological styles while sharing sardonic factual news. For example, fake news may quote factual evidence in an indecent context to support a false statement. Thus, obtainable, handcrafted, and data-specific textual features are commonly insufficient for fake news recognition. Supplementary facts are also useful to increase uncovering data, such as knowledge base and user social network engagements. Second, exploiting this supporting evidence leads to additional dangerous challenges where data quality itself is a challenge. Fake news is usually correlated to newly developing, time-critical happenings, which may not have been appropriately tested by existing knowledge bases due to the absence of supporting evidence or statements. Besides, users' social actions with fake news yield large, incomplete, unstructured, and noisy data. Effective methods to distinguish trustworthy users, mine useful post features, and exploit network exchanges are an open area of research and need further investigations. In this article, we present an overview of fake news recognition and discuss encouraging research directions. The fundamental inspirations of this investigation are described as follows:

- To do better investigation in fake news detection research,
- To understand and review existing fake news recognition methods under social networking medias like Facebook, Twitter, Whatsapp, etc., and to understand basic state of the art in fake news detection techniques,
- Fake news detection on social media is still in the premature stage of progress, and there are still many interesting problems that need additional investigations. It is necessary to deliberate impending research directions that can increase fake news recognition and justification competences.

Fig. 9.1 lists the statistics of the social media website utilizers by January 2019. From the figure, around 2.27 billion users are using Facebook to share their status and news, YouTube is in second place with 1.9 billion users, WhatsApp in next place with 1.5 billion users, after that Facebook Messenger with 1.3 billion users, WeChat with 1.08 billion, Instagram 1 billion, QQ with 803 million users, and QZone, Douyin/Tik Tok, and Sino Weibo have 531, 500, and 446 million users, respectively.

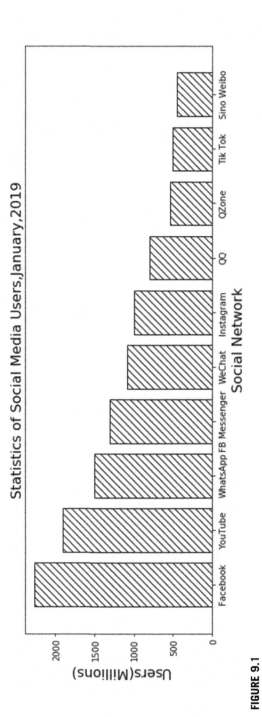

FIGURE 9.1

Social media users' statistics.

To expand the investigation in fake news detection on SM, an in-depth survey was done in two aspects of fake news detection, termed characterization and detection. Initially, we explain the history of fake news identification from genuine news using theories and properties from sensibility in addition to societal studies; then detection approaches will be discussed. Our main contributions for the book chapter include these:

- study the state of art of the existing algorithms aiming to distinguish fake news versus real news in social networks
- elaborated discussion on definitions and terminology used in fake news
- collection, analysis, and visualization of fake news and related dissemination on social networks
- proposing a method to detect fake news
- comparative analysis with existing methods

2. Literature review

The research in fake news detection has become an emergent topic in the latest social media data sharing analysis. In general, fake news detection approaches generally are divided into two categories: by news content and by social context [1]. In the first category, content-based approaches, phonological or linguistic or photographic features are extracted. Phonological features, such as syntactic features and lexical features, capture definite writing styles and amazing headlines that frequently arise in fake news contents [2], while photographic features are used to recognize fake pictures that are purposely produced or internment specific characteristics for pictures in fake news [3]. Models that exploit the news content-based features can be classified into knowledge and the style based. In knowledge-based type, external sources are used to test the genuineness of statements in news subjects [4,5], whereas in style-based type traces the manipulation in inscription style, such as deception [6] and nonobjectivity [2]. As for social context-based methods, they integrate features from user profiles, contents posted, and social networks. User profiles can be used to quantify the users' characteristics and trustworthiness [7]. The features extracted from the users' posts signify the users' social responses, such as attitude [8]. Network features are taken out by building a specific social network, such as diffusion networks [9] or cooccurrence networks [10]. The social context models are categorized into either stance based or propagation based. The stance-based models utilize the users' feelings or thoughts toward the news to deduce news exactness [11], while propagation-based models apply propagation methods to model unique outlines of information propagation [11−20]. The aforementioned methods are all supervised approaches that mainly focus on extracting effective features and use them to construct supervised learning frameworks.

Symbolization used: Table 9.1 describes the symbols used in the proposed method along with their meaning.

Table 9.1 Symbols used in MLSTM and their description.

Symbol	Description	Symbol	Description
S_{t-1}	Previous cell status	W	Weight vector
h_{t-1}	Previous cell hidden state	b	Bias value
f_t	Forget gate	P	Probability
i_t	Information gate	X	Vector multiplication
σ	Sigmoid function	o_t	Output gate
X_t	Current input	s_t	Current cell status
+	Pointwise addition	h_t	Current hidden status

3. Existing fake news detection techniques

This section describes some of the existing algorithms developed by various authors to automate the process of fake news recognition. Here is a detailed description of the most frequently preferred algorithms:

1. **Naïve Bayes:** This classifier algorithm was developed to get a baseline accuracy rate for the considered dataset. Specifically, the scikit-learn implementation of Gaussian naïve Bayes mechanism was done to classify fake news. This is one of the modest methodologies suitable to do the classification through a probabilistic approach, with the conjecture that all features are provisionally independent given the class label. As with the other representations, Doc2Vec embedding was used, which is to be discussed in Section 4, the naïve Bayes rule basically uses the Bayes' theorem parameter estimation for naïve Bayes model to use maximum likelihood. The advantage of this technique is that it has a need for only a minor quantity of training data to calculate the parameters.

$$P(x|c) = \frac{P(x|c)P(c)}{P(x)} \tag{9.1}$$

2. **Support vector machine (SVM):** In 1963, Vladimir N. Vapnik and Alexey Ya. Chervonenkis [21] proposed the original technique. But, the main limitation of this proposed technique, which can do linear classification, which is not suitable for many other practical problems. Bernhard E. Boser et al. [22] proposed a technique in 1992 using the kernel trick that enables the SVM to classify the datasets for nonlinear classification. That creates a very influential SVM.

Radial basis function kernel is used in our project experimentation; the reason to use this kernel is that two Doc2Vec feature vectors will be close to each other if their corresponding documents are alike, so the distance calculated through the kernel function should still symbolize the original distance. Since the radial basis function correctly represents the relationship we aspire for and it is a common kernel for

SVM, we use the system introduced in Ref. [23] to implement the SVM. The chief idea of the SVM is to discriminate different classes of data by the broadest "street." This goal can be denoted as the optimization problem we finally solve by using the convex optimization tools provided by a Python package called CVXOPT.

3. **Long short-term memory (LSTM):** It is a kind of recurrent neural network (RNN), where output produced in one step will become the input to another step; similarly the output from the final step serves as input in the present step. It has the capability to handle the long-term dependencies of an RNN problem, where the RNN cannot foresee the word stockpiled in the long-term memory but can give more accurate likelihoods from recent information. RNN does not give an effectual performance with the increase in gap length. LSTM has the feature of retaining information for a long time and is mainly used for processing, guessing, and classifying on the basis of time-series data.

$$i_t = \sigma(W_{xi}X_t + W_{hi}h_{t-1} + b_i) \tag{9.2}$$

$$f_t = \sigma(W_{xf}X_t + W_{hf}h_{t-1} + b_f) \tag{9.3}$$

$$g_t = \tanh(W_{xg}X_t + W_{hg}h_{t-1} + b_g) \tag{9.4}$$

$$o_t = \sigma(W_{xo}X_t + W_{ho}h_{t-1} + b_o) \tag{9.5}$$

$$S_t = S_{t-1} * f_t + i_t * g_t \tag{9.6}$$

$$h_t = \tanh(\pi S_t * o_t) \tag{9.7}$$

4. **Feed-forward neural network:** Mainly, two possible feed-forward network models exist: one is through Tensorflow and another one is through Keras. Neural networks are used in modern natural language processing (NLP) applications [24,25], in contrast to existing approaches that primarily concentrate on linear models such as SVMs and logistic regression. The internal arrangement of these two technique implementations uses three hidden layers. In the Tensorflow implementation, all layers had around 300 neurons each, wherein the Keras implementation used different sizes of layers, where layers of size 256, 256, and 80 were intermixed with dropout layers to avoid overfitting. In the implementation of the activation function, the rectified linear unit (ReLU) was chosen, which has been found to execute well in NLP applications.

This has a fixed-size input $x \in R^{300}$

$$h1 = \text{ReLU}(W_1 x + b_1) \tag{9.8}$$

$$h2 = \text{ReLU}(W_2 h_1 + b_2) \tag{9.9}$$

$$y = \text{Logits}(W_3 h_2 + b_3) \tag{9.10}$$

4. **Proposed method:** This section describes the detailed phases used in fake news and genuine news recognition. Fig. 9.2, shows the block diagram of fake news detection using MLSTM technique.

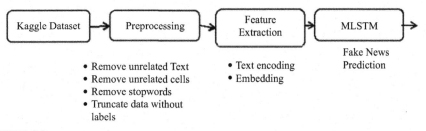

FIGURE 9.2

Block diagram of fake news detection using MLSTM technique.

It mainly comprises preprocessing and feature extraction along with Kaggle dataset classification through modified long short-term memory (MLSTM). In detail, the description is as follows:

3.1 Dataset description

The news datasets used were collected from Kaggle [26], divided into training and testing dataset. During experimentation, 70% of the dataset is used as the training part and the remaining 30% is preferred as the testing dataset. In the Kaggle data set, we observed the existence of around 16,600 rows of records from various news articles on the internet. We had to do quite a little preprocessing of the news articles [27] to train our models. Table 9.2 explains the detailed description of the attributes in the Kaggle dataset. Fig. 9.3 shows the in-depth word cloud representations of the training as well as test datasets used for experimentation.

Table 9.2 Attributes description of Kaggle fake news dataset.

Sl-No	Attribute name	Description
1	Id	Unique identification number assigned to the article
2	Title	The heading of a news article, which is a small text aiming to catch the attention of readers that describes the theme of the article
3	Author	Author of the news article
4	Text	The article body part that exclusively elaborates the depth details of the title that was mentioned in the title attribute. This is specially emphasized and shapes the perspective of the publisher
5	Label	This attribute is intended to mark the article as potentially defective by assigning two values 1: erratic fake and 0: reliable genuine

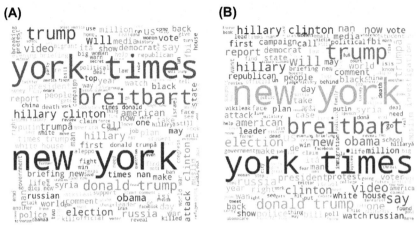

Wordcloud of Training Dataset   Wordcloud of Testing Dataset

FIGURE 9.3

Word cloud representation of training and testing Kaggle datasets.

3.2 Preprocessing

The preprocessing phase aims for clean text data by performing the following:

- remove unrelated text;
- remove empty cells;
- remove stop words;
- truncate data without labels.

There are numerous techniques commonly used to transform text data into a form that is suitable for modeling. The data preprocessing steps mentioned before like the removal of unnecessary information will be applied to headlines as well as the news articles. After this phase, the finely tuned news article will be forwarded to the feature extraction phase.

3.3 Feature extraction

After removing stop words, removing special characters and punctuation, and converting text into lowercase, the resulting text is passed into the feature extraction phase. This phase produces a words vector of each article. In proposed architecture, Doc2Vec model is used with naïve Bayes and SVM, and text encoding and embedding mechanisms were used with LSTM and MLSTM.

This list of words separated by commas was taken out from preprocessing, which can be sent into the Doc2Vec procedure [28] to generate a length of 300 size embedding vector for each article. Doc2Vec is a methodology proposed by Quoke [29] in 2014 developed from the existing Word2Vec technique, which generates word vectors in general. Word2Vec characterizes documents by linking the vectors of the individual words, but in doing, it misses all word information order. Doc2Vec magnifies the Word2Vec technique by including a parameter "document vector" to the output representation, which encompasses some information about the document in total and permits the model to learn some evidence about word order. Safeguarding of word existence order information makes Doc2Vec useful for our application, as we are aiming to detect subtle differences between text documents. The proposed method has used a paragraph-based vector framework, where each paragraph is mapped onto an exclusive vector called a paragraph vector, denoted by D matrix, which is a single-column matrix, and each word in a paragraph is mapped onto a vector called a word vector, represented by W, which is a column vector. The paragraph and W are merged or concatenated to anticipate the next word in a connection context. More formally, the only change in this model compared to the word vector framework is in shown Fig. 9.4, where h is constructed from W and D. It acts as a memory unit, responsible for remembering the missing context from the current statement from the paragraph. For this reason, our time and again model is called paragraph vectors distributed memory model.

The D and W are trained using stochastic gradient descent, and the gradient is obtained via backpropagation. In every step, a fixed-length textual context is selected from a random paragraph, the error gradient is calculated and then the gradient is used to update the parameters in the model. Afterward, the prediction will be initiated, during which there is a need to perform an interpretation to figure out the paragraph vector for a newly selected paragraph, which is also gained by gradient descent. In this phase, for the parameters for the rest of the model, we fixed out W and the softmax weights. Suppose that there are N paragraphs present in the corpus, M words in the language, and we want to learn paragraph vectors, then each paragraph and the words are mapped to p dimensions and q dimensions, respectively, and the model has an overall $N \times p + M \times q$ parameters by eliminating the softmax parameters. Even though the number of parameters can be large (N is large), the changes during training are typically sparse and thus effectual.

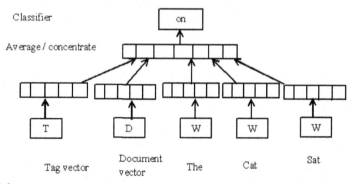

FIGURE 9.4

A framework for learning paragraph vector.

3.4 MLSTM

The sequence of the word was an important feature for the MLSTM unit, which makes us use the Doc2Vec in the preprocessing phase because the transformation of the entire document into one vector changes the order of words, so sometimes we may lose the order in the information. To avoid that, we have used a process called word embedding in the place of word sequencing. Afterward, we remove all the special characters and symbols from the document. Then the frequency of each word is counted and appended in our training dataset to prepare the most frequent 5000 words. Later, each word is assigned with a unique identification number. Assuming that these 5000 are the possible common words that cover most of the text, we may lose little information but need to convert the string into a list of integers. Since the MLSTM unit needs a fixed input vector length, we shorten lists longer than 500 words because more than half of the news considered in the experimentation is longer than 500 words. Then for that list with less than 500 words, we padded 0's at the beginning of the list to discriminate longer text from the shorter text for processing. We also remove the news with only a few words since they do not carry adequate information for training dataset preparation. By doing this, we can convert the original news text string into a fixed-length integer vector while preserving the word order information. Finally, we use word embedding to transfer each and every word ID to a 32-dimension vector. The word embedding process trains each word vector based on word resemblance or similarity. When the two words frequently appear together in the news text, they are considered to be more alike, and the distance between corresponding vectors is small. The output collected from the feature extraction after the completion of embedding is then forwarded to the modified MLSTM unit to train the model.

RNN, where the output from the last step is lent as input in the present step, can tackle the difficult long-term dependencies of RNN in which the technique is unable to predict the word deposited in the long-term memory but can give more accurate predictions from the recent information. Fig. 9.5 shows a single memory cell in MLSTM architecture, aiming to improve the performance over LSTM, where the following two changes were made to standard LSTM architecture,

1. In LSTM, the amount of past information to delete and the amount of new information to be added was decided separately, which may lead to missing some information. But, in MLSTM, the amount of information to be deleted is estimated based on the amount of new information that needs to be added. Hence, the architecture first calculates the new information to be added is based on the information from gate output and then calculates the amount to forget.
2. Secondly, to increase accuracy and to lessen the error rate, tanh function from hidden state ($h(t)$) is removed.

After the aforementioned modifications, the modified LSTM architecture is not only works more accurately, it also reduces total execution time. The modified equations are these:

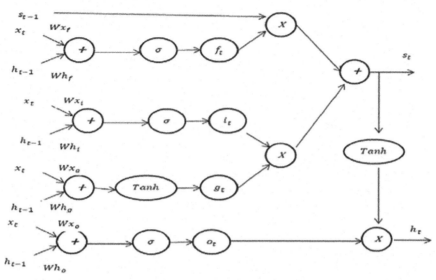

FIGURE 9.5

MLSTM single-cell forward-step architecture.

$$i_t = \sigma(W_{xi}X_t + W_{hi}h_{t-1} + b_i) \qquad (9.11)$$

$$f_t = 1 - i \qquad (9.12)$$

$$g_t = \tanh(W_{xg}X_t + W_{hg}h_{t-1} + b_g) \qquad (9.13)$$

$$o_t = \sigma(W_{xo}X_t + W_{ho}h_{t-1} + b_o) \qquad (9.14)$$

$$S_t = S_{t-1} * f_t + i_t * g_t \qquad (9.15)$$

$$h_t = S_t * o_t \qquad (9.16)$$

MLSTMs are capable of providing more performance in time intervals above 1000 time steps even in the presence of noisy words and incomprehensible input word sequences, without loss of information [30]. The architecture imposes constant error flow through internal states of a storage unit known as the memory cell. This memory cell has connectivity with four gates: forget gate, input gate, information gate, and output gate.

3.4.1 Basic operations
The basic operations used in the gates used in the modified MLSTM are sigmoid, tanh, pointwise addition, and pointwise multiplication.

1. Tanh activation: This operation helps to regulate the values that are passing through the network. This activation function arranges the values between -1 and 1, thus affecting the output of the neural network.

2. Sigmoid function: The gates present in the LSTM memory cell use this sigmoid activation function. This activation function is similar to the tanh activation function. But, the difference is that it ensures the values between 0 and 1 instead of between −1 and 1. This will be helpful for forgetting data, because any value when multiplied by 0 results in 0, causing values to be "forgotten." The value of the number, which is multiplied by 1, stays the same, so that number is "kept." Hence the network can learn to decide which data is unimportant, and it can be forgotten by keeping the important data.

3. Pointwise addition: This operator performs the summation of the values piece-wise. This is used to merge or combine current input and old memory.

4. Pointwise multiplication: This operator is used to multiply the old memory with a vector that is close to 0, which means to forget most of the old memory. The forget gate value is made equal to 1 when we need to remember the old memory.

3.4.2 Memory cell building blocks

From Fig. 9.5, we can observe that the memory cell in MLSTM comprises the following gates:

1. **Forget gate**: The first stage in the MLSTM memory cell is the forget gate, in which the sigmoid function results in values ranging from 0 to 1 decide how much information of the previous hidden state and current input should be remembered. Forget gates are necessary to the performance of MLSTM because the network does not necessarily require remembering everything that has happened in the past. The information transferred is subtracted from 1, and then the resulting values are mapped between 0 and 1 and come out as the result. If the value is closer to 0, it means to gate has to forget; otherwise, if it is closer to 1, it means to remember. In fake news recognition, when we are moving from one news article text to the next in the testing dataset, we can decide whether to forget all of the information related to the old news article or remember it.

2. **Input gate**: This gate is used to update the cell state, initially a previously hidden state and current input state to be input into a sigmoid function with ranges of values between 0 and 1. Then, pass the hidden state and current input to the tanh function to squish values between −1 and 1 to help regulate the network. Afterwards, the tanh output is multiplied with the sigmoid output. The sigmoid output decides the information that is important to keep from the tanh output.

3. **Updating the cell memory**: At this point, we have determined what to forget and what to input, but we have not changed the memory cell state. Now, we should likely have abundant information for estimating cell state. Primarily, the cell state gets pointwise values that are multiplied by the forget vector. Here, there is a possibility of dropping values in the cell state when it gets multiplied by values that are near 0. Then, a pointwise operation is performed on the input gate output values, and the cell state is updated to new values so the neural network can find an appropriate new cell state.

4. **Output gate**: To determine memory cell output, a sigmoid function is applied on the previously hidden state and current input, and the result is multiplied by

tanh; then, it is forwarded to the new memory cell (this will make the values between −1 and 1). In the fake news identification model example, we want to output information that will help predict the next sequence of notes. This gate decides the state of the next hidden state, which is done by remembering the hidden state that contains information about the previous inputs and its states. The hidden state is usually intended for state predictions. Initially, we need to pass the previous hidden state value and the current input state value into a sigmoid function. Then, the modified cell state value coming from the sigmoid function is passed to the tanh function. Finally, it multiplies the tanh output state with the sigmoid output gate values to decide which information will be carried by the hidden state. The new cell state and hidden state are forwarded to the next time step, which is shown in Fig. 9.6.

After understanding the detailed operations used and the functionality of the individual gates, we can summarize that MLSTM is a RNN whose arrangement is like LSTM cell blocks. These cells have gate components like input, forget, and output gates. We have our new word/sequence values Xi and their weights connected to the previous state output s_{t-1} from the cell. The initial step combines the received input values through the tanh layer. In the next step, this input is crossed through an input gate. This is a layer of sigmoid-activated nodes, where the node-generated output is

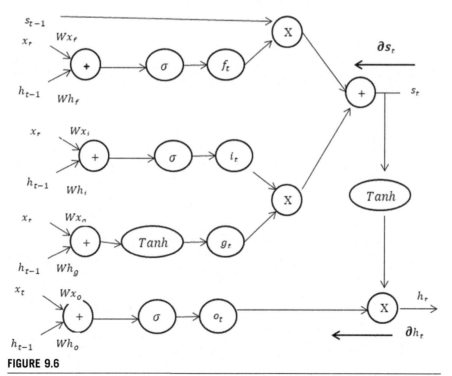

FIGURE 9.6

MLSTM single memory cell backward propagation structure.

multiplied by the compressed input, and these input gate sigmoids can make the decision to remember or to "kill off" any elements in the input vector that are not essential. A sigmoid function produces values between 1 and 0, so the weights associating the input to these nodes can be trained to produce output values of 0 for the closest value to zero; otherwise the input value will be passed to the next step when the output is closest to 1. The next step is an internal state/forget gate loop where the flow of data through this cell is continuous. An additional operation is used in the place of a multiplication operation, which will help to reduce the risk of vanishing gradients. However, the forget gate takes the responsibility to control the recurrence loop. Finally the output value comes out from the output gate.

With these equations, Eqs. (9.11)–(9.16), we can update the states as the feed forward from the initial step 1 to step T.

To make the prediction, a linear model over the hidden state can be added along with the output likelihood with the soft matrix function:

$$Z_t = \text{softmatrix}(W_{hz}h_t + b_z) \tag{9.17}$$

At time t the reality is we try to minimize the least square, so for the top layer classification, we need to take the derivative w.r.t respectively.

$$\partial Z_t = y_t - z_t \tag{9.18}$$

$$\partial W_{hz} = \sum_t h_t \partial Z_t \tag{9.19}$$

$$\partial h_T = W_{hz}\partial Z_T \tag{9.20}$$

Here, we are considering the gradients w.r.t. at the time step T. For any time step t, the gradient will be slightly different, so the back propagation of the MLSTM at the time step t is given as follows:

$$\partial o_t = \tanh(S_t)\partial h_t \tag{9.21}$$

$$\partial S_t = \left(1 - \tanh(S_t)^2\right)o_t\partial h_t \tag{9.22}$$

$$\partial f_t = S_{t-1}\partial S_t \tag{9.23}$$

$$\partial S_{t-1} + \; = f_t\partial S_t \tag{9.24}$$

$$\partial i_t = g_t * \partial S_t \tag{9.25}$$

$$\partial g_t = i_t * \partial S_t \tag{9.26}$$

Where the gradient w.r.t $\tanh(t)$ in Eq. (9.21) can be derived according to the equation in the appendix.

Further, the backpropagation activation functions over the whole sequence in MLSTM are given as shown next:

$$\partial W_{x0} = \sum_t o_t(1 - o_t)x_t\partial o_t \tag{9.27}$$

$$\partial W_{xi} = \sum_t i_t(1 - i_t)x_t\partial i_t \tag{9.28}$$

$$\partial W_{xf} = \sum_t f_t(1-f_t)x_t \partial f_t \tag{9.29}$$

$$\partial W_{xs} = \sum_t \left(1-(g_t)^2\right)x_t \partial g_t \tag{9.30}$$

$$\partial W_{h0} = \sum_t o_t(1-o_t)h_{t-1}\partial o_t \tag{9.31}$$

$$\partial W_{hi} = \sum_t i_t(1-i_t)h_{t-1}\partial i_t \tag{9.32}$$

$$\partial W_{hf} = \sum_t f_t(1-f_t)h_{t-1}\partial f_t \tag{9.33}$$

$$\partial W_{hs} = \sum_t \left(1-(g_t)^2\right)h_{t-1}\partial g_t \tag{9.34}$$

Finally the hidden state at current step $t-1$ is calculated as

$$\partial h_{t-1} = o_t(1-o_t)W_{h0}\partial o_t + i_t(1-i_t)W_{hi}\partial i_t + f_t(1-f_t)W_{hf}\partial f_t + \left(1-(g_t)^2\right)W_{hs}\partial g_t \tag{9.35}$$

We can observe that there is a possibility of error in each time step that will be increased as it is passing backward. Fig. 9.7 shows the arrangement of the unfolding of the memory cell unit, considering that we have the least mean square objective function:

$$\xi(x,\theta) = \min\sum_t \frac{1}{2}(\partial z_t)^2 = \min\sum_t \frac{1}{2}(y_t - z_t)^2 \tag{9.36}$$

Where $\theta = \{W_{h0}, W_{hi}, W_{hf}, W_{hs}, W_{x0}, W_{xi}, W_{xf}, W_{xs}\}$ by ignoring the bias function. For easy understanding, $\xi(x,\theta)$ is represented as $\xi(t) = \frac{1}{2}(y_t - z_t)^2$.

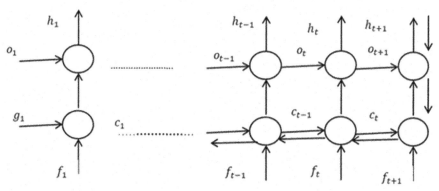

FIGURE 9.7

Unfolding of the memory unit of MLSTM to understand error propagation.

At the time step T, find the derivative of $\xi(T)$ w.r.t S_T:

$$\frac{\partial \xi(T)}{\partial S_T} = \frac{\partial \xi(T)}{\partial h_T} \frac{\partial h_T}{\partial S_T} \tag{9.37}$$

Also, find the derivative of ξ at time step $T-1$ w.r.t S_{T-1}:

$$\frac{\partial \xi(T-1)}{\partial S_{T-1}} = \frac{\partial \xi(T-1)}{\partial h_{T-1}} \frac{\partial h_{T-1}}{\partial S_{T-1}} \tag{9.38}$$

From Fig. 9.7, we can observe that error propagation is not only the propagation from current state but also from previous state, so the final gradient is given as follows:

$$\frac{\partial \xi(T-1)}{\partial S_{T-1}} = \frac{\partial \xi(T-1)}{\partial S_{T-1}} + \frac{\partial \xi(T)}{\partial S_{T-1}} \tag{9.39}$$

$$= \frac{\partial \xi(T-1)}{\partial h_{T-1}} \frac{\partial h_{T-1}}{\partial S_{T-1}} + \frac{\partial \xi(T)}{\partial h_T} \frac{\partial h_T}{\partial S_T} \frac{\partial S_T}{\partial S_{T-1}} \tag{9.40}$$

4. Performance analysis

Research prefers that how many parameters is used in the performance analysis of classification algorithms. The confusion matrix is one among them. We have used this in performance comparison analysis of various fake news classification algorithms. It is a table comprising two rows and columns, representing two types of fake news correct prediction and two types of fake news incorrect prediction. Table 9.3 shows the confusion matrix of fake news prediction.

From the confusion matrix, the following details will be furnished:

- True positive (TP): The algorithm-predicted value is matched with the reality that news is fake. We can conclude that the algorithm has correctly classified and news shared in the social network is fake.
- False negative (FN): The predicted output is a false negative, where news is incorrectly classified that the news shared is negative even though the news is genuine.
- True negative (TN): Predicted output is a true negative when the algorithm-predicted value is matched with the reality that news is genuine. We can conclude that the algorithm has appropriately classified.

Table 9.3 Confusion matrix for fake news prediction.

	Predicted news is fake (I)	Predicted news is genuine (0)
News is fake (1)	TP	FN
News is genuine (0)	FP	TN

- False positive (FP): News is inaccurately classified that shared news is genuine news even though it is fake news.
- Negative (N): A 0 value is used to represent a negative case, which means the news is genuine.
- Positive (P): A value of 1 is used to represent a positive case, which means the news is fake.

Once the confusion matrix was constituted, the performance of the data classification algorithms was compared by doing the comparative analysis using parameters classification accuracy, classification error, sensitivity or recall, specificity, precision, and Matthew Correlation Coefficient (MCC). These parameters are determined as follows:

$$\text{Accuracy} = \frac{\alpha}{\beta} \times 100\% \tag{9.41}$$

$$\text{Classification Error} = \frac{\mu}{\beta} \times 100\% \tag{9.42}$$

$$\text{Recall} = \frac{TP}{\Omega} \times 100\% \tag{9.43}$$

$$\text{Precision} = \frac{TP}{\eth} \times 100\% \tag{9.44}$$

$$\text{MCC} = \frac{\text{subtract}(TP * TN, FP * FN)}{\sqrt{\mu * \alpha * \Omega * \eth}} \times 100\% \tag{9.45}$$

where $\beta = \text{sum}(TP, TN, FP, FN)$, $\alpha = \text{sum}(TP, TN)$, $\mu = \text{sum}(FP, FN)$, $\Omega = \text{sum}(TP, FN)$ and $\eth = \text{sum}(TP, FP)$

5. Experimentation and results

To conduct the experimentation on the proposed technique, a dataset is taken from Kaggle (the description of which is shown in Table 9.4). For training the news article content a maximum threshold value is set on the sequence length as a hyperparameter. The sample template of the dataset along with their labeling where 0 means unreliable and 1 means reliable news is shown in Table 9.5. Development of the algorithm was done in the Anaconda environment on Intel(R), Core i5, CPU @3.10 GHz, where we implemented our modified MLSTM framework in Keras [31], following a pattern composed of seven layers, as described in Section 4. In the feature extraction, phase

Table 9.4 Statistics of the dataset.

Dataset statistics	News articles
Training set size	20,799
Testing set size	4153

Table 9.5 Sample news article and headlines.

ID	Title	Author		Label
0	House Dem Aide: We Didn't Even See Flynn	Dairell Lucus	House Dem Aide: We Didn't Even See Comey's Letter Until Jason Chaffetz	1
1	Hillary Clinton, Big Woman on Campus	Daniel J. Flynn	Ever get the feeling your life circles the roundabout rather than heads in	0
2	The Major Potential Impact of a Corporate Tax Overhaul	Neil Irwin	The United States' system for taxing businesses is a mess	1
3	Russian Researchers Discover Secret Nazi Military Base Treasure Hunter in the Arctic	Armando Flavio	The mystery surrounding the Third Reich and Nazi Germany is still a subject of debate between many observers. Some believe that Nazi Germany, under the control of Adolf Hitler, possessed supernatural powers and lately employed pseudoscience during the 1933–1945 period	1
4	Chuck Todd Did Donald Trump a Political Favor	Jeff Poor	Wednesday after Donald Trump's press conference at Trump Tower in New York City	0

encoder and embedded steps have experimented with the following statistics, MLSTMs with two layers with 100 cells at each layer in encoder phase and 200 dimensional word embeddings with input lexemes of 5000 words. Similarly, for the decoder, we have used the same arrangement used in the encoder phase. The word embeddings are arbitrarily initialized, and they are learned along with the network. We train the network for 400 epochs with a batch size equal to 64 (the number of training examples utilized in one iteration) using stochastic gradient descent as optimization for loss function, employing the ReLU as activation function at the convolution layer and sigmoid as an activation function in the output layer, as shown in Table 9.6. All these features were normalized before applying various classification algorithms

Table 9.6 Hyperparameter tuning performance.

Parameter name	Possible range values	Chosen value
Activation function	ReLU or tanh or softmax	Softmax
DataSet dimension	Min: 32, max: 256	64
Dropout level	Min: 0, max: 1	0.1
Period value	Min: 50, max: 200	50
L2 penalty	{0,1,0.01,0.001,0.0001}	0.0001

(KNN, ANN, SVM, and Decision tree). The comparative analysis in terms of confusion matrix is shown in Fig. 9.8 and Table 9.7. From these, we can observe that the proposed technique is performing better.

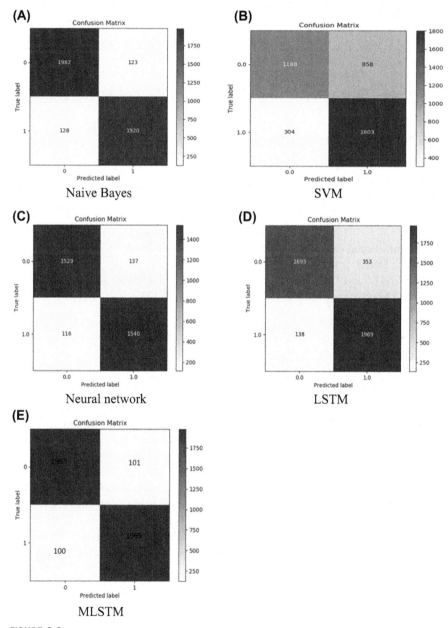

FIGURE 9.8

Confusion matrix of different classification techniques.

Table 9.7 Comparative analysis of existing techniques.

Method	Accuracy (%)	Classification error (%)	Precision (%)	Recall (%)	MCC (%)
Naïve Bayes	72.02	27.98	58.06	79.62	68.81
SVM	88.18	11.82	82.75	92.46	120.15
Neural network	92.38	7.62	91.78	92.95	85.55
LSTM	93.96	6.04	94.16	93.93	138.62
MLSTM	95.11	4.89	95.19	95.06	142.30

Table 9.8 LSTM versus MLSTM error propagation.

Method↓	Error propagation
LSTM	$0.631e^{-6}$
MLSTM	$0.169e^{-6}$

Computational complexity: The time complexity of both LSTM and MLSTM is equal to $O(KH + KCS + HI + CSI)$. In-depth comparison of error propagation between LSTM and MLSTM algorithms is shown in Table 9.8, and it is observed that MLSTM error propagation is less compared to LSTM.

6. Conclusion and future work

A complete, quality classifier technique will incorporate many different features beyond the vectors corresponding to the words in the news article. For fake news detection, we can add the sources of the news, includes associated URLs, the news category (e.g., science, business, politics, or sports), publication medium (blog, print, or social media), country, or publication year. This chapter has presented a mechanism termed MLSTM that can distinguish fake news from genuine news. Its functionality effectiveness is tested using the Kaggle dataset, and we observed that its performance is good. Also, a comparative analysis was done and showed that MLSTM performance is better than other existing techniques using neural network, LSTM, naïve Bayes, and SVM models.

Future extension: The work can be extended by adding the following functionalities:

- Combine multiple classifiers to achieve better performance.
- Propose a method to identify the sources of the news.
- Develop a method to search the news from the web to crosscheck the authenticity of the news shared.

Appendix:

1. Gradients related to $f(t) - \tanh(t)$

Take the derivative of the $f(x)$ w.r.t t

$$\frac{\partial f(t)}{\partial t} = \frac{\partial \tanh(t)}{\partial t} = \frac{\partial \frac{\sinh(t)}{\cosh(t)}}{\partial t} \tag{9.46}$$

$$= \frac{\frac{\partial \sinh(t)}{\partial t}\cosh(t) - \sinh(t)\frac{\partial \cosh(t)}{\partial t}}{\cosh(t)^2} \tag{9.47}$$

$$= \frac{\cosh(t)^2 - \sinh(t)^2}{\cosh(t)^2} = \left(1 - \tanh(t)^2\right) \tag{9.48}$$

2. Gradients related to $f(t) = \sigma(t)$

$$\partial f(t) = \sigma(t)(1 - \sigma(t)) \tag{9.49}$$

Where Sigmoid function $\sigma(t) = \frac{1}{1+e^{-t}}$

References

[1] H. Allcott, M. Gentzkow, Social media and fake news in the 2016 election, J. Econ. Perspect. 31 (2) (2017) 211−236. https://web.stanford.edu/oegentzkow/research/fake news.pdf.

[2] M. Potthast, J. Kiesel, K. Reinartz, J. Bevendorff, B. Stein, A Stylometric Inquiry into Hyperpartisan and Fake News, 2017 arXiv preprint arXiv:1702.05638.

[3] M. Gupta, P. Zhao, J. Han, Evaluating event credibility on twitter, in: Proceedings of the 2012 SIAM International Conference on Data Mining, SIAM, 2012, pp. 153−164.

[4] M. Amr, N. Wanas, Web-based statistical fact checking of textual documents, in: Proceedings of the 2nd International Workshop on Search and Mining User-Generated Contents, ACM, 2010, pp. 103−110.

[5] K. Wu, S. Yang, K.Q. Zhu, False rumors detection on sina weibo by propagation structures, in: Data Engineering (ICDE), 2015 IEEE 31stInternational Conference on, IEEE, 2015, pp. 651−662.

[6] V.L. Rubin, On deception and deception detection: content analysis of computer-mediated stated beliefs, Proc. Assoc. Inf. Sci. Technol. 47 (1) (2010) 1−10.

[7] J. Abernethy, O. Chapelle, C. Castillo, Graph regularization methods for web spam detection, Mach. Learn. 81 (2) (2010) 207−225.

[8] Z. Jin, J. Cao, Y. Zhang, J. Luo, News verification by exploiting conflicting social viewpoints in microblogs, in: AAAI, 2016, pp. 2972−2978.

[9] S. Kwon, M. Cha, K. Jung, W. Chen, et al., Prominent features of rumor propagation in online social media, in: International Conference on Data Mining, IEEE, 2013.

[10] N. Ruchansky, S. Seo, Y. Liu, Csi: a hybrid deep model for fake news detection, in: Proceedings of the 2017 ACM on Conference on Information and Knowledge Management, ACM, 2017, pp. 797−806.

[11] B.D. Deebak, F. Al-Turjman, A novel community-based trust aware recommender systems for big data cloud service networks, Sustainable Cities and Soc. (2020) 102274.

[12] L. Li, B. Qin, W. Ren, T. Liu, Document representation and feature combination for deceptive spam review detection, Neurocomputing 254 (2017) 33−41.

[13] C. Buntain, J. Golbeck, Automatically identifying fake news in popular twitter threads, in: Smart Cloud (SmartCloud), 2017 IEEE International Conference on. IEEE, 2017, pp. 208−215.

[14] D. Esteves, A.J. Reddy, P. Chawla, J. Lehmann, Belittling the Source: Trustworthiness Indicators to Obfuscate Fake News on the Web, 2018 arXiv preprint arXiv:1809.00494 (2018).

[15] S. Chaudhry, H. Alhakami, A. Baz, F. Al-Turjman, Securing demand response management: A certificate based authentication scheme for smart grid access control, IEEE Access 8 (1) (2020) 101235−101243.

[16] C. Shao, G. Luca Ciampaglia, O. Varol, A. Flammini, F. Menczer, The Spread of Fake News by Social Bots, 2017 arXiv preprint arXiv:1707.07592 (2017).

[17] C. Silverman, This Analysis Shows How Viral Fake Election News stories Outperformed Real News on Facebook, BuzzFeed News 16, 2016.

[18] F. Al-Turjman, Intelligence and security in big 5G-oriented IoNT: An overview, Future Gener. Comput. Syst. 102 (1) (2020) 357−368.

[19] Y. Wang, F. Ma, Z. Jin, Y. Yuan, G. Xun, K. Jha, S. Lu, J. Gao, EANN: event adversarial neural networks for multi-modal fake news detection, in: Proceedings of the 24th ACM SIGKDD International Conference on Knowledge Discovery & Data Mining, ACM, 2018, pp. 849−857.

[20] D. Zhang, L. Zhou, J. Luo Kehoe, I.Y. Kilic, What online reviewer behaviors really matter? Effects of verbal and nonverbal behaviors on detection of fake online reviews, J. Manag. Inf. Syst. 33 (2) (2016) 456−481.

[21] V.N. Vapnik, A.Y. Lerner, Pattern recognition using generalized portraits, Autom. Remote Control 24 (1992) 709−715. 1963. Russian original: В.Н. ВаШникњ АћЯћ Лернерњ Узнавание образов Ири Иомощи обобщенныч Иортретовњ Автоматика и телемечаника ЫЭ ЋЯЦ ЋЪыЯЭЦ щщЭ−щъШћ Тжд нрзёќ змак артзвкд сублзттдг нм ЫЯ Гдвдлбдр ЪыЯЫћ

[22] B.E. Boser, I.M. Guyon, V.N. Vapnik, A training algorithm for optimal margin classifiers, in: D. Haussler (Ed.), Proceedings of the 5th Annual Workshop on Computational Learning Theory (COLT '92), ACM Press, New York, NY, USA, Pittsburgh, PA, USA, July 27-29, 1992, pp. 144−152.

[23] C.M. Bishop, Pattern Recognition and Machine Learning, 2016. http://users.isr.ist.utl.pt./wurmd/Livros/school/Bishop%20%20Pattern%20Recognition%20And%20Machine%20Learning%20-%20Springer%20%202006.pdf.

[24] Y. Goldberg, A Primer on Neural Network Models for Natural Language Processing, 2015. https://arxiv.org/pdf/1510.00726.pdf.

[25] P. Tahmasebi, A. Hezarkhani, Application of a modular feedforward neural network for grade estimation, Nat. Resour. Res. 20 (1) (21 January 2011) 25−32, https://doi.org/10.1007/s11053-011-9135-3.

[26] Datasets, Kaggle, February 2018. https://www.kaggle.com/c/fake-news/data.

[27] Source Code Repository, GitHub, 2018. https://github.com/FakeNewsDetection/FakeBuster.

[28] M. Zuckerberg, Facebook Post, November 2016. https://www.facebook.com/zuck/posts/10103253901916271.

[29] Q. Le, T. Mikolov, Distributed representations of sentences and documents, in: Proceedings of the 31st International Conference on Machine Learning, PMLR 32, 2014, pp. 1188–1196.

[30] H. Li, G. Fei, S. Wang, B. Liu, W. Shao, A. Mukherjee, J. Shao, Bimodal distribution and co-bursting in review spam detection, in: Proceedings of the 26th International Conference on World Wide Web, International World Wide Web Conferences Steering Committee, 2017a, pp. 1063–1072.

[31] A. Ray, P. Agarwal, C.K. Maurya, G.B. Dasgupta, Creative tagline generation framework for product advertisement, IBM J. Res. Dev. 63 (1) (2019) 6:1–6:10. https://doi.org/10.1147/JRD.2019.2893900.

Security aspects and UAVs in Socialized Regions

10

Deepanshu Srivastava[1], S. Rakesh kumar[1], N. Gayathri[1], Fadi Al-Turjman[2]
[1]*School of Computing Science and Engineering, Galgotias University, Greater Noida, Uttar Pradesh, India;* [2]*Research Center for AI and IoT, Near East University, Nicosia, Mersin, Turkey*

1. Introduction

Unmanned aerial vehicles (UAVs) were initially used for military purposes for arming forces against rival territory. They are also used in various civil applications, which will be discussed in further sections. Applications like farming and agriculture, public safety and convenience, environment protection measures, and traffic control in efficient ways are uses in many sectors in smart cities. Nowadays, how these UAVs are involved in smart cities is one of the emerging areas. They improve interactivity and quality of life, which allows the management of sectors such as agriculture, pollution, traffic, energy sources, government, and water.

To make a successful implementation, projects like multiple interconnected systems and sensors require mobile and stationary UAVs. A UAV can be well described as an aircraft or aerial vehicle that can work and fly without a pilot inside, and it is designed in a variety of scales from small to full-sized aircraft. It can provide useful methods and instruments to work and maintain a smart city and implement its goals to provide an efficient infrastructure.

2. UAVs: a brief introduction

UAVs have a wide range of applications: safety, scientific research, and other applications [1]. The architecture components consist of a processor system, monitoring system, and landing control system in which the processor system consists of a navigation function. This market is still growing and solving a new problem every day.

The most important target of a smart and modern city is to give the best infrastructure and provide services at minimum cost. Europeans describe smart cities as the use of new methods and technology and resources such as internet of things (IOT) in more intelligent and integrated ways, referring to a modern style of living. Moreover, a smart city is the city that can achieve the objectives and necessities of the future as well as an excellent link up with Information and Communication Technology (ICT).

After the global financial recession, interest of people has emerged for the smart cities evolution since the population of the world is increasing; it is assumed to be doubled or more by 2060 [1].

The design of such smart cities requires a full contribution to its goals. UAVs, as in Fig. 10.1 have a wide range of functions that can help in growing smart cities. These applications can handle traffic flow and detecting floods by wireless sensors. They are greatly used in detecting natural disasters. This work aims at detailing the use of UAVs in smart cities.

The main factors that provide initiatives for a smart city are given in Fig. 10.2 as follows:

1. Management is required for the accomplishment of goals that are necessary for the effective and efficient working of a smart city.

FIGURE 10.1

UAV (unmanned aerial vehicle).

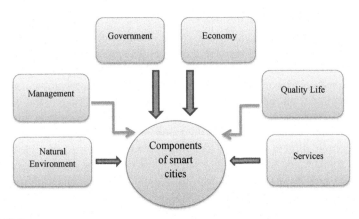

FIGURE 10.2

Components of smart city.

2. Organization is required for the smart technology applied to critical components and services. Computing in a new generation refers to integrated hardware and software that can provide more intelligent decision-making patterns.

3. Government is required to process the construction and exchange of information that helps to achieve the goals. Data exchange and leadership is required for the government of smart city. The development of policies that help to create urban development characterizes urban issues that are nontechnical.

4. Economy is one of the most important topics in the smart and modern city, as with the high degree of economy that would result in high job creation, high development, and large-scale production in cities.

5. Natural environment and infrastructure: ICT is required for the smart city development with its availability and performance. It is used to increase sustainability and resource management [1].

Opportunities: Many opportunities exist that support the formation of smart cities and will be discussed in this section. These will be beneficial to a smart city in terms of city growth and development as well as strong financial stability.

1. Geo-surveying activities are one of the new UAV civil applications used for geospatial surveying. Better optimization of data provides wireless networks as sensors as the main component of any such system using UAVs. Thus it provides a base to the technology in smart cites, resulting in high performance and low power consumption. Due to reliability, integration of such technology is possible.

2. Traffic management: Efficiency of safety and traffic management is one of the serious concerns for the smart cities, not only smart cities but any type of city and area. Involvement of UAVs in police activities has been supported by US and other top-level agencies. It can help smart cities for a better place of living at a larger extent.

3. Agriculture management: It can be helpful in fertilizing the crops by dropping fertilizer, as in Fig. 10.3 and water with the help of UAV systems. It can monitor the growth of the crops and environment by its surveillance camera and can measure any harmful condition.

4. Urban security: The opportunity of UAVs in a smart city is security management (urban security). The main motive is for efficient data flow and allows managing big public events successfully with full technical coverage.

5. BIG data: For smart infrastructures, BIG data processing is required with different requirements, such as support with GIS, combination of modeling and simulation, and time data processing.

Challenges: There are two types of challenges that will be discussed in this section, commercial or business and technical.

1. Privacy business challenge: Many people think UAVs are an invasion to their privacy, which created a large debate in the United Kingdom due to wrong utilization of UAVs. As they are available for general purpose use, one can spy on someone's activities.

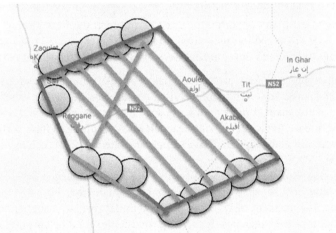

FIGURE 10.3

UAV dropping fertilizer and its working ground length.

2. Cost business challenge: They can be expensive due to deployment, technical issues, and integration combination of systems. Specific designing of UAVs is also expensive for proper functioning.
3. License business challenge: Licenses are important for using a flying UAV in any country. First, it should be registered in that location. A flying UAV may affect the airplane routes, so a country must provide some rules for the UAV deployment.
4. Adoption: UAVs are a challenge for several companies to run due to certain aspects in meeting their business requirements and can be costly in term of resources. Amazon has introduced their own drone called Prime Air, which can deliver the product to a customer in 30 min [1,6].

2.1 Technical challenges

1. UAVs provide accurate and stable measurement of field strength and development of efficient engines and a gyro-stabilized level for high pixel resolution images.

2. Wireless sensors can be used for correct and good operations of UAVs. They can be used for a traffic management system.
3. The UAV can work during both days and nights in all weather conditions, whether they are suitable or not, as execution of a flying drone (height and flight) path over extended period of time is required.
4. Development of a network center for any team member to control the UAV and get the images and real-time information is a challenge.

3. UAV using IOT in smart cities

The latest UAVs can be implemented in various ways, ground and aerial, controlled in two techniques, manually as well as automated. They are available in miniature and large scale. The most common and suitable way is to use the aerial UAV because of its relevant speed and independence of traffic jams. More processing power can be provided to UAVs by the latest technology.

3.1 IOT platforms

Designing any smart UAV requires four aspects: data, device, service, and connectivity. Its physical component comprises a cloud server, which stores collected data, performs coordination tasks, communicates with UAVs, and hosts the operator interface. Operator, client, and UAV are controlled by the platforms and operators.

Controlling hardware and software can be integrated as a part of IOT structure and can be implemented for communications with UAVs, as connection between device and cloud server for interdevice communication.

Access should be provided to UAVs by human operator, energy resources should be managed, and aircraft stations should be available for recharging the UAVs, where replacements are also available. Multiple air stations should be available throughout the city, so they can provide complete maintenance at any facility of battery recharge. Batteries should be provided separately from UAVs for charging purpose as a measure of safety, as in Fig. 10.4. Battery type used is lithium-polymer. Its exchange is possible and charging is time is greater [4,13].

3.2 Case study

Multiple applications of UAV can be executed for smart cities, such as agriculture, telecommunication, defense, resource, and environment analysis. It can further used as a signal booster and node for IOT and sensor networks. Deployment of UAVs and cost will be reduced compared to traditional aircraft and transport facilities.

Farming sector: One of the growing sector for UAVs is the agriculture sector, as depicted in Fig. 10.5. UAVs can be used for monitoring crops and distribution of chemical over the crops. They can also monitor the current condition of fields and stock available, as well as fertilizers.

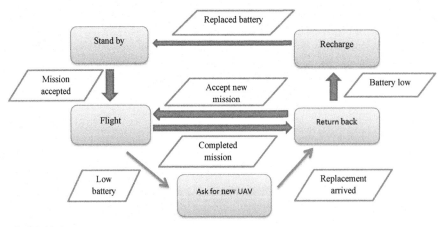

FIGURE 10.4

Function of a UAV.

FIGURE 10.5

Agricultural UAV.

One of the popular replacements for airplanes is the latter application because of its various benefits: cost is reduced, lower carbon footprint, less fertilizer and pesticide loss, and noise pollution is reduced.

The whole setup is done by using GPS coordinates when designing the path and then covers the area with the based speed.

Complete paths and coverage area targets can be completed in segments by using multiple UAvs. Easy visual control of a UAV is provided, and then combining multiple UAVs into one unit works while flying on a path.

Another one of the growing fields for the implementation of UAVs in a smart city is video and camera surveillance. It increases security and reduces crimes to some extent [11].

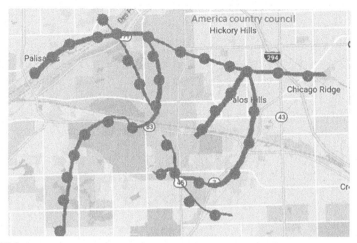

FIGURE 10.6

CCTV camera locations.

According to the Urban Institute Justice Policy Center, these fixed surveillance cameras reduced crime around 15%–40%, depending on the area. The whole system consists of nearly 650 cameras, as depicted in Fig. 10.6 [10].

It is cost efficient as well as effective because when crimes are executed, less money is used automatically in safeguarding citizens [4,7].

4. UAV transportation system

ITS (intelligent transport system) is one of the most important components of the smart city project. Next-stage ITS technologies such as autonomous vehicles are at the end of completion for large-scale deployment. The testing phase of the latest technologies has been started in many countries, and effective measures are taken to regulate future technologies, as these services will enable many other services and applications will be processed further [2].

Military forces have been using UAVs or drones for several years, in which there is an increment in use of these drones in the fields of surveillance, security, farming, and agriculture. Giants like Amazon, Wal-Mart, and many others are using this technique to deliver shipments to customers. Companies in China are trying to deliver at the rate of 600 parcels every day to customers with the help of these UAVs. Automated vehicles are necessary for the complete automation of transportation in which many other services like traffic police need to be automated.

UAVs (ITS) can provide efficient traffic rules, and efficient traffic information to the user can be provided by intelligent traffic management (ITS). Some of the great applications that can be provided by ITS UAVs are flying dynamic traffic signals and flying speed cameras, as depicted in Fig. 10.7.

FIGURE 10.7

Connected by drone server that can save from danger ahead.

4.1 ITS for smart cities

Smart cities provide an improved quality of life for their residents by providing automated systems, whether they are semiautomated or fully automated, such as transportation systems and security systems. Future smart cities and everything will be connected to the internet. It has been deployed and provided in many areas, like Spain, where all transportation systems have been connected via a common control room where data is gathered and processed to make their services more efficient, where users can get the details of their transportation services [16].

These can improve the road safety measures as well as transportation routes, which can give users an efficient pathway from their home to work destination. They can exchange real-time information to make better safety services and make better road measures provided with sensors, which makes them more time efficient as the user will not be required to wait for the transportation services. It is one of the great and important steps toward the automation of transport systems [2,7,12].

4.2 Applications

UAVs can be used for autonomous driving. They can improve traffic, give better safety measurements, and provide more comfort to the driver of the vehicle. Some problems that still need to be solved consist of limiting energy, signal

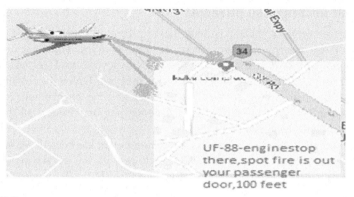

FIGURE 10.8

Details about the fire accident provided by the UAV drone.

processing, and advanced processor functions. There is a great potential for the utilization of UAV drones in the future.

1. Accident reporting UAV: When an accident occurs on the road, the lives of humans totally depend upon the rescue team and how fast a team can reach the accident; sometimes due to inefficient conditions, they can be delayed.

   As shown in Fig. 10.8, a rescue team needs to opt for flying parameters such as rescue helicopters or rescue jets, which might be costly and not good for the cities. In such scenarios, UAV drones are an optimal solution that can help the help rescue team to reach the accident on time.

   The selection of a UAV can be made based the number of UAVs present at that time and the distance between the UAV and accident. It helps to give a brief and detailed report about the accident location that can be remotely handled. As it has a provision to supply products and items, it can carry a first aid kit and provide it at the accident location.

2. Police UAV: Nowadays, traffic police are equipped with the latest technologies. CCTV is the most common technique to implement traffic safety rules. If someone exceeds the speed limit, then it will be recorded by the CCTV. By the time people get to know how to tackle these traffic rules, they adjust their speed according to CCTV camera locations. Technology like embedded speed cameras on police vehicles can also be used in the UAVs.

   UAVs can also fly over a road and catch any vehicle for speeding or breaking the traffic rules, as depicted in Fig. 10.9. They can also help condition surveys and counting vehicles on roads. The communication between UAVs needs to be maintained every time [8].

4.3 Cyber security and privacies

When there is a network, there exist securities to protect it from various attacks, and to provide those security measures, there are some policies. These are one of the most important provisions for the security and privacy of the UAV smart cities. When all the things will be interconnected by the means of UAVs, then all the details will be collected onto storage devices. A situation may arise where details of vehicle location and information are leaked for the wrong purpose. Hence, all these data must be protected from third parties.

Homomorphic encryption is one of the techniques that has been developed using different approaches. It is grouped into generations with the underlying approach categorized into first generation FHE, second generation Fully Harmonic Encryption (FHE), and third generation FHE.

A user at one side sends the information encrypted with the public key by the function to encrypt the formation; then at the other side, another user decrypts the data using his private key. FHEs are becoming a useful technology with aspect to UAVs. The most important thing is to provide flexible and efficient security to UAV-enabled ITS software and applications.

One of the consequences that may affect UAV deployment is limitation of energy, i.e., the battery of the UAV, which is usually not more than 30 min. If we change battery type or make it dependent on solar source, that can be somewhat effective. We need to create careful security and privacy of the network of UAVs, which is another challenging task [6].

5. UAV issues in cyber security and public safety

As the growth of UAVs is increasing, it is expected that they will play a major role in smart cities. As their growth brings introduction and production of new technology, they also bring the concerns and challenges related to security, especially cyber security. There are many benefits attached with the use of UAV drones, but they can also be used for malicious activities. The main aim of this section is to see different problems and challenges related to UAV drones in the near future.

According to the rules developed by the FAA, they required a UAS weight of less than 55 lbs and more than 0.55 lbs in their systems. If not taken according to the rules

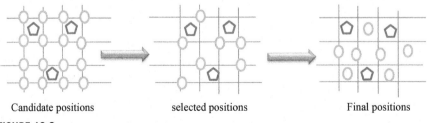

Candidate positions   selected positions   Final positions

FIGURE 10.9

UAV deployment algorithm.

of the FAA, the owner can face criminal penalties and problems. Rules include a height of flying of 400 ft with no obstacles around and maintaining the line of sight. A campaign was started by the FAA, "know before you fly," which explains all the UAV safety regulations. The FAA can also detect, investigate, and stop the illegal and unauthorized use of UAV drones [5].

Now, we will look at some popular small UAVs that are commercially available and compare them to see the differences between them: 3D robotics solo drone, Phantom 2 vision drone, and Parrot AR drone 2.0 [14].

1. 3D robotics solo drone: It is world's first smart aerial vehicle with 1 GHz Arch (Linux) computers, and these drones, as in Fig. 10.10 are manufactured by 3D robotics with a camera attached and enabled that also gives HD video recording as output through an application in phones. It is more secure compared to other drones because it has strong password protection. Cyber hacking of this drone might be a difficult task.
2. Phantom 2 vision drone: This drone, as in Fig. 10.11 is used for aerial view photography and was founded in 2006 by DJI. A mobile phone is connected with the drone that can connect up to 300 m, providing resolution up to 14

FIGURE 10.10

3D Robotics solo drone.

FIGURE 10.11

Phantom 2 vision drone.

FIGURE 10.12

Parrot AR drone 2.0.

megapixels. Live video is available at DJI vision app provided by 2.4–2.488 GHz band. It also has two other Linux systems: one is used for encoding using the IP address and other one is used for images and recording purposes.

3. Parrot AR drone 2.0 was first introduced in 2010 and was built by a French company. Later, its version 2.0, which came in 2012, was given by CES Las Vegas. According to specification of this drone, as in Fig. 10.12 it consists of a 1-GHz 32-bit ARM cortex A8 processor. A busy box is also attached to it, which is a Linux machine that can run across various platforms like Linux and Android. It can be managed by applications, whether on android or IOS, for storing images and video recordings. It has a tool kit present called AR drone toolkit. The main aim of this type of drone is for surveillance and spying over unidentified objects. They can also be controlled by programming languages, e.g., Python, JS, node.js, and are very popular [5,7].

Attacks on drones: All type of drones are controlled, and communication between drone and server is done by the use of WIFI using IEEE standards. They can be hacked because of no encryption, an IBM researcher said.

Skyjack is a software that can control other UAVs within range and creates many drones as an army of drones, as in Fig. 10.13. It can detect all networks, especially wireless, and deactivate them from the drone. This issue can be solved by protected WIFI, which provides password login. Since all UAVs discussed earlier have WIFI, they can create a wireless network and connect them. These connecting drones can be managed with the WIFI Pineapple device, as it helps in performing automated task [15].

Spoofing attack through GPS is a type of a cyber-attack executed on the UAV drones. Communication is established between them, including incoming signals from GPS, for presence of drones and communication from the ground. GPS can navigate a drone because of no encryption available, so they can easily be spoofed. Loss of a drone is possibly due to cyber-attack on its GPS, i.e., GPS spoofing attack by transmitting fake signals to control the drone. In this attack, a transmitter is used to provide the wrong signals [5,9].

SYSTEM FAILURE

FIGURE 10.13

System failure of UAV.

Satellites transmit correct signals to a drone to move it on its desired path. When an attacker is near the drone, then this attack is possible. It can be easily attacked.

Cyber-attack tools: We have seen many advantages of drones, but like a coin, it also has two sides: drones can also be used in malicious ways.

Sniffing signal using VPN: WIFI Pineapple at a drone can be used to sniff the signals. There are multiple ways a device can be accessed, e.g., ethernet cable connected on a laptop or connecting a device with a wireless network. As these drone fly at heights far away, these methods may be not possible. A VPN (virtual private network) is created with WIFI Pineapple and the devices, so it can access any device from anywhere with the help of the internet. It creates a 2014-bit symmetric key that will be exchanged between client and server by key exchange method, i.e., Diffie Hellman key exchange.

Unauthorized drones can be executed in any country for surveillance recording and can used to execute many types attacks, as discussed earlier, which can harm human society to a great extent. Uses of these drones are increasing as many industries have started using these drones because of availability at low cost. So it is difficult to identify unauthorized drones [5].

One way a UAV can be brought down is using another UAV. Hacking of a drone through cyber processing with the help of another drone is a possible technique due to use of a drone using a WIFI network. Parrot drone can be directed using WIFI Pineapple. Now, this drone acts as a slave. When shell commands are executed, this slave drone is under control of WIFI Pineapple.

6. UAV in smart cities: Dubai

UAVs are mostly used in civil applications, public safety, and traffic control. The main aim of any smart city is to provide effective and efficient services at low cost. After the global recession, interest in smart cities is increasing. There will be many more benefits on the establishment of a smart city. Transforming Dubai into a smart city provides opportunities that can be beneficial. This work aims to provide use of UAV in smart cities by considering Dubai, as in Fig. 10.14. In the beginning, UAVs were used for military purposes, which had a bad effect on society. Afterward, when used for civil works, the image of UAVs has improved in society [1,3].

UAVs also made a remarkable image in NEPAL by protecting wildlife, which to some extent helped stop a wildlife crisis. Some of the affecting factors are these:

1. Government: It helps to achieve the goals and objectives by exchanging valuable information according to provided rules and standards.
2. Community: It helps to allow the citizens of the city to participate in management; if their work is good, then they have the opportunity to participate in activity that influences success of events.
3. Environment: The goal is to enhance natural resource management. Protection of these natural resources and increasing sustainability is extremely important.
4. Economy: One of the most important factors in building a smart city and one with a high economy is to lead to more business creation, more job creation, work development, and positive improvement in production processes.
5. Policy context: It is important to understand the proper use of these systems in effective ways, which helps to solve various issues in development processes.

FIGURE 10.14

Flying UAV in Dubai.

Table 10.1 Airspace classification.

Airspace classes	Airspace equivalents	Changes
A	Positive control area (PCA)	None
B	Terminal control area (TCA)	VFR: clear of clouds
C	Airport radar service (ARSA)	None
D	Airport traffic area (ATA)	Upper limits of 2500 AGI
	Control area (CZ)	
E	General controlled airspace	Visibility cloud clearance
G	Uncontrolled airspace	Visibility cloud clearance

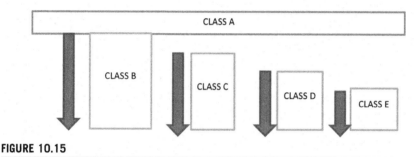

FIGURE 10.15

Airspace classes.

Table 10.1 focuses more on class A, as in Fig. 10.15, where most planes fly. Then the airspace near airports called class B, class C, and class D small cities are described [1,3].

Opportunities that are beneficial to Dubai to make a more effective and efficient smart city are as follows.

1. Environmental management: UAVs can be used to a large extent to drop fertilizer and water on crops. They can also monitor the emission of harmful gases and elements.
2. Surveying activities: The proper optimization of data by sensor networks are main parts that can be used by any UAVs. This requires real-time integrated software that provides high performance and low power consumption. It creates a large opportunity for fire management applications in which their use will be beneficial. Geographic surveying is an additional application that is helpful in analyzing the environment.
3. Disaster management: UAVs help in disaster management, assisting authorities at that situation in handling those conditions. It can provide assistance in surveying the affected area properly as they can reach areas where humans cannot.

4. Traffic management: With the help of UAVs, safety and efficiency of traffic in cities can be managed. Top political federations have also supported the use of UAVs in various countries. Addition of applications in mobile and secure networks can lead to creating safe, smart cities.

6.1 Challenges

1. Cost: One of the major factors is cost; development of UAVs can be quite expensive because of system issues and technical maintenance. The testing phase itself can be costly if it shows some error in the drone.
2. Business techniques: Running a UAV is a difficult task for large companies because they require additional resources on the other side; it will also give a good gain to the company.
3. Privacy: Many people think that it may be an invasion to their privacy, so it might be possible that they do not approve of UAVs monitoring cities. If used in wrong ways, UAVs may create issues in ethics.
4. Technical: Sometimes, due to a technical fault in the UAV, a problem may exist when all communication with drone is lost.
5. Middleware: Establishment of middleware to provide smooth operation in a UAV is a challenging task [3].

References

[1] F. Mohammed, A. Idries, N. Mohamed, J. Al-Jaroodi, I. &Jawhar, May). UAVs for smart cities: opportunities and challenges, in: 2014 International Conference on Unmanned Aircraft Systems (ICUAS), IEEE, 2014, pp. 267–273.
[2] H. Menouar, I. Guvenc, K. Akkaya, A.S. Uluagac, A. Kadri, A. &Tuncer, UAV-enabled intelligent transportation systems for the smart city: applications and challenges, IEEE Commun. Mag. 55 (3) (2017) 22–28.
[3] F. Mohammed, A. Idries, N. Mohamed, J. Al-Jaroodi, I. &Jawhar, March). Opportunities and challenges of using UAVs for dubai smart city, in: 2014 6th International Conference on New Technologies, Mobility and Security (NTMS), IEEE, 2014, pp. 1–4.
[4] A. Giyenko, Y. &Im Cho, Intelligent UAV in smart cities using IoT, in: 2016 16th International Conference on Control, Automation and Systems (ICCAS), IEEE, October 2016, pp. 207–210.
[5] E. Vattapparamban, İ. Güvenç, A.İ. Yurekli, K. Akkaya, S. &Uluağaç, Drones for smart cities: issues in cybersecurity, privacy, and public safety, in: 2016 International Wireless Communications and Mobile Computing Conference (IWCMC), IEEE, September 2016, pp. 216–221.
[6] D. He, S. Chan, M. &Guizani, Communication security of unmanned aerial vehicles, IEEE Wireless Commun. 24 (4) (2016) 134–139.
[7] A. Gharaibeh, M.A. Salahuddin, S.J. Hussini, A. Khreishah, I. Khalil, M. Guizani, A. Al-Fuqaha, Smart cities: a survey on data management, security, and enabling technologies, IEEE Commun. Sur. Tutorial. 19 (4) (2017) 2456–2501.

[8] Y. Mehmood, F. Ahmad, I. Yaqoob, A. Adnane, M. Imran, S. &Guizani, Internet-of-things-based smart cities: recent advances and challenges, IEEE Commun. Mag. 55 (9) (2017) 16—24.

[9] A.S. Elmaghraby, M.M. &Losavio, Cyber security challenges in Smart Cities: safety, security and privacy, J. Adv. Res. 5 (4) (2014) 491—497.

[10] F. Al-Turjman, L.J. Poncha, S. Alturjman, L. Mostarda, Enhanced deployment strategy for the 5G drone-BS using artificial intelligence, IEEE Access 7 (1) (2019) 75999—76008.

[11] F. Al-Turjman, A Novel Approach for Drones Positioning in Mission Critical Applications, Wiley Transactions on Emerging Telecommunications Technologies, 2019, https://doi.org/10.1002/ett.3603.

[12] Z. Ullah, F. Al-Turjman, L. Mostarda, Cognition in UAV-aided 5G and beyond communications: a survey, IEEE Trans. Cognitive Commun. Network. (2020), https://doi.org/10.1109/TCCN.2020.2968311.

[13] F. Al-Turjman, M. Abujubbeh, A. Malekoo, L. Mostarda, UAVs assessment in software-defined IoT networks: an overview, Elsevier Comp. Commun. J. 150 (15) (2020) 519—536.

[14] F. Al-Turjman, S. Alturjman, 5G/IoT-Enabled UAVs for multimedia delivery in industry-oriented applications, Springer's Multi. Tool. Applicat. J. (2018), https://doi.org/10.1007/s11042-018-6288-7.

[15] F. Al-Turjman, H. Zahmatkesh, I. Aloqily, R. Dabol, Optimized unmanned aerial vehicles deployment for static and mobile targets monitoring, Elsevier Comp. Communications Journal 149 (2020) 27—35.

[16] S. Muthuramalingam, A. Bharathi, N. Gayathri, R. Sathiyaraj, B. Balamurugan, IoT based intelligent transportation system (IoT-ITS) for global perspective: a case study, in: Internet of Things and Big Data Analytics for Smart Generation, Springer, Cham, 2019, pp. 279—300.

Index

Printed in the United States
By Bookmasters